Map of Pennsylvania, Showing the Areas Surveyed in 1874, 1875, 1876, 1877 & 1878.

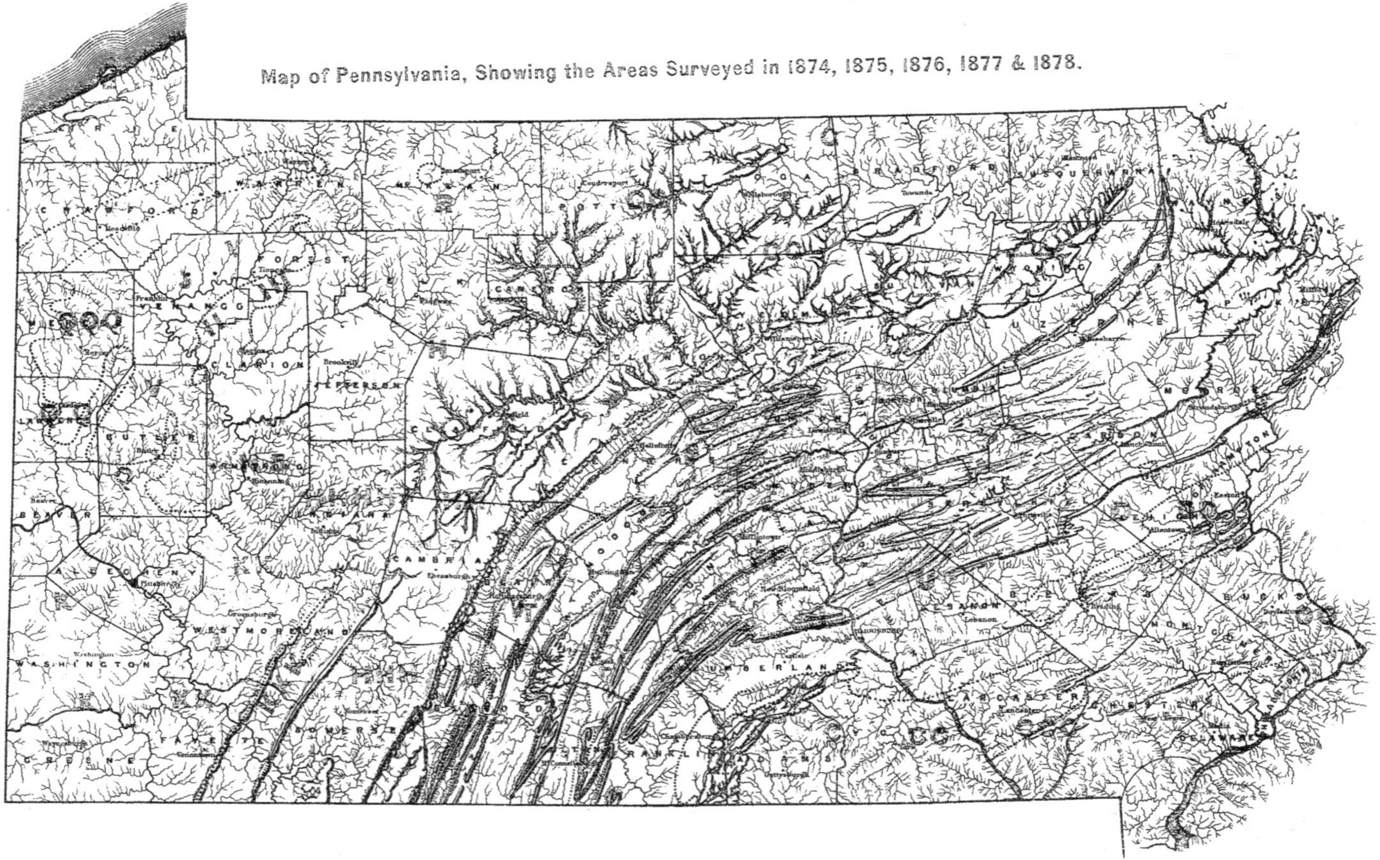

SECOND GEOLOGICAL SURVEY OF PENNSYLVANIA:
1874-'5-'6.

HISTORICAL SKETCH

OF

GEOLOGICAL EXPLORATIONS

IN

PENNSYLVANIA AND OTHER STATES

BY

J. P. LESLEY.

WITH AN APPENDIX CONTAINING

THE

ANNUAL REPORTS OF THE STATE GEOLOGIST TO THE BOARD OF COMMISSIONERS.

HARRISBURG:
PUBLISHED BY THE BOARD OF COMMISSIONERS
FOR THE SECOND GEOLOGICAL SURVEY.
1876.

LANE S. HART,
State Printer and Binder,

BOARD OF COMMISSIONERS.

His Excellency, JOHN F. HARTRANFT, *Governor*,
and *ex-officio* President of the Board, Harrisburg.

ARIO PARDEE, - - - - - - Hazleton.
WILLIAM A. INGHAM, - - - - Philadelphia.
HENRY S. ECKERT, - - - - - Reading.
HENRY M'CORMICK, - - - - - Harrisburg.
JAMES MACFARLANE,- - - - - Towanda.
JOHN B. PEARSE, - - - - Philadelphia.
ROBERT V. WILSON, M. D., - - Clearfield.
HON. DANIEL J. MORRELL, - - Johnstown.
HENRY W. OLIVER, - - - - - Pittsburg.
SAMUEL Q. BROWN, - - - - - Pleasantville.

SECRETARY OF THE BOARD

JOHN B. PEARSE, - - - - - Philadelphia

STATE GEOLOGIST.

PETER LESLEY, - - - - - - Philadelphia.

PREFACE BY THE AUTHOR.

The continuation of my Historical Sketch of Geological Explorations in Pennsylvania, begun in this volume, has been postponed from time to time on account of the pressure of official duties.

Sixteen volumes of reports of progress have been issued from the press; five more are printed and nearly ready to issue; another is in press; and seven others will be published during the current summer and autumn; making twenty-nine volumes in all.

To secure accuracy and completeness as far as possible for these published reports of my colleagues; to preserve harmony of nomenclature; to prevent the presentation of conflicting hypotheses; to furnish direction, advice, and assistance when demanded by parties in the field; to draw and color, and superintend the drawing and correction of illustrations; to arrange and write classified indexes for all the volumes; and to read proof of every page going through the press, has been a ceaseless, anxious occupation, apart from the inevitable correspondence of the Survey.

I have therefore no leisure to attempt a revision and correction of these pages now reprinted in a special edition, by order of the Legislature, for the service of the members. But I cannot let escape this opportunity for publishing a letter which I have received from Mr. Wm. B. Rogers, jr., of Philadelphia, to show that if I have done injustice to any one whose labors in science fall within the scope of my historical sketch, it has not been intentionally done.

PHILADELPHIA, *June 28, 1878.* J. P. LESLEY.

Letter of Mr. W. B. Rogers, jr.

1000 WALNUT STREET,
PHILADELPHIA, *September 11, 1877.*

Prof. J. P. LESLEY,
State Geologist:

DEAR SIR: An impulse, not unnatural or unworthy, as I believe, prompts, and must be my apology for this communication.

I have recently read, with a great deal of interest, your "Historical Sketch of Geological Explorations in Pennsylvania, &c.," (1876,) and it is to this subject I beg leave to ask your attention.

You have devoted many pages of narrative to each year in the progress of the First Geological Survey of Pennsylvania, from 1836 to 1842, and have, no doubt, fairly assigned credit to the several assistants engaged upon it. Your intimate knowledge of the mass of material contributed to the final report, and your own work in its preparation, fitted you to become the historian of the first Survey. Your acquaintance with the details of the operations of the *Revision* of the first survey ends with the close of its first year, 1851.

It is here, early summer of 1852, that my personal relations to the Survey began, and they were continued without break until the last sheet of the final report had issued from the Edinburgh press.

In those six years of constant devotion, I became only less well acquainted than yourself with the previous surveys, and in the revised work I was at home in every detail, and upon many points the factotum. My anxious and harrassed uncle trusted to my naturally ready memory to supply the defects of his own. My pen, my pencil, all my acquired information and my zeal, were at his service by day and by night. All of these were accepted at their just value, and I had the consciousness, which still rests with me, of labor well or at least faithfully performed.

Many years have since intervened, and my course in life has taken a different direction, and yet, even now, I read with pride the language of so keen a critic as yourself in commendation of geological cross-sections and cuts, chiefly illustrating the Anthracite coal chapters, which you attribute to the "genius" of Whelpley, or "the skill" of McKinley, but which were the production of my own hand and "crow-quill" pen, from originals in the pencilled field drawings of one, sometimes of both of my uncles, but more frequently results of the toilsome labor of myself and my faithful attendant, Patrick Daly, whose judgment, almost amounting to intuition in the tracing of coal outcrops, was at that day, and, I believe still is, regarded as unrivaled.

Your impression that the First Survey closed with 1851 is very widely in error. In the years 1852, 1853, and 1854 the entire productive domain of the anthracite coal was mapped in the field, not by copying Whelpley or anybody else, excepting as to the framework of outlying mountains, but continuing the surveys of 1851 in the pen and pencil work of Dalson, on the ground, upon "planchettes," previously prepared by Mr. Poole, of plottings of highways and railways, and cross-sections more or less numerous, which myself and party spent much time in running.

A portion of the Wyoming basin was actually triangulated from a base measured on the "flats" opposite Wilkes Barre. Of the text, too, it may be said that nearly every page of practical detail was written in the region which it describes. The lines of flexure and of outcrop were laid down on the field maps, and the descriptions followed them without the slightest reference to the early Surveys. At Boston, and in Philadelphia, these chapters were supplemented by others having a more general, practical, or scientific bearing.

More or less rapid visits were made to various parts of the State, and notes were taken of the more progressive portions of the bituminous coal field and of some of the iron districts, for the purpose of freshening the old material. The entire Geology of the S. E. counties was revised and re-written by Prof. Rogers, and illustrated by sections executed in the closet, and very often in the field, by the writer of these lines.

It is not my purpose to write the history of the years after 1851, until the appearance of the Final Report. I have meant only to illustrate, in a simple way, the errors into which you have been led by your devotion to the good name and fame of your talented associates of the First Survey. Some of these are, no doubt, due to inadvertence in the haste of writing your resumé of the contents of the several chapters of the Final Report, *e. g.*, credit to Lehman for pictorial illustrations which bear the imprint of "Dalson," as the "Newkirk colliery," and the overturned outcrop, the "Tuscarora breaker," (then just built,) the "Roaring Brook Falls, &c."

Upon a review of the facts stated, it would seem proper to remark that your statement, (p. 127,) that "the First Survey of Pennsylvania may be truly said to close in 1851" needs considerable qualification, and, that "superficial revision in the field, and editorial labor in the cabinet," cannot reasonably account for the long interval of six years prior to 1858.

Am I not also justified in asserting that the language of the late State Geologist in reference to myself is, presumably, true, and that the single notice with which you honor me, (foot note, page 126,) suggests a false inference. If I am not justified, I find it somewhat difficult to account for those years of my life which have made me in some respects the being I am. You say that the Survey closed in 1851; that "W. B. Rogers, jr., was not seen by those engaged in this Survey of 1851 at all"—*de non apparentibus et non existentibus eadem est ratio.*

I wish to express my satisfaction with the opinion expressed by you (p. 107) in reference to the conduct of the first survey, and the explanation thereof, (p. 72 top.) No detail of narrative less explicit than that into which you have, *con amore*, entered, could do justice to the subject of individual merit, but "consistency and completeness" in the *result* is, after all, the first duty of the editor.

In the "struggle for existence," I have long been withdrawn from the ambitious career of a "man of science," but association and training compel an interest in the progress of every good scientific work. I note with pleasure the elaborate completeness of the "Second Survey" over which you preside, and the judgment with which, in directing it to practical ends, you secure popular appreciation and support, while your scientific intelligence is free for a more general study of the field of exploration.

In freedom and with candor I have written this long letter, and thanking you for your patience, if you have read thus far,

I remain with respect,

Very sincerely yours,

WM. B. ROGERS, Jr.

CHAPTER I.

EARLY OBSERVATIONS OF THE GEOLOGY OF PENNSYLVANIA.

Although the neighborhood of Philadelphia was explored by amateur mineralogists as early as 1820, and those large and beautiful private cabinets of minerals began then to be collected which afford rich materials for Dr. Genth's Report on the Mineralogy of Pennsylvania, to be published, this winter, as part of the first fruits of the Second Geological Survey of the State, it was not until about the year 1830 that intelligent eyes were cast upon vegetable and animal fossils, and curious minds undertook the problems of structural geology. The impulse came from New Haven, Boston, Troy and New York; but the discipline was home-made in the halls of the American Philosophical Society and of the Academy of Natural Sciences. Philadelphia had long been the chief centre of natural science on this side of the Atlantic; so that, when the science of Von Buch and DeBeaumont, Sedgwick and Murchison reached Philadelphia, it found the companions and disciples of Bartram and Wilson, McClure and Say, DaSerra, George Ord, and Isaac Lea, ready to receive it and able to advance its progress on American ground.

The early geological memoirs of this country, published in the Transactions of the American Philosophical Society at Philadelphia, in the Transactions of the American Academy at Boston, and subsequently in the Transactions of the Connecticut Academy of Arts and Sciences at Hartford, the Proceedings and Journal of the Academy of Natural Sciences at Philadelphia, and one or two scientific magazines, chiefly Professor Silliman's American Journal of Science and Art at New Haven, were of the crudest nature, but indicative of a widespread desire to observe useful facts in nature and a child-like ignorance of their deeper meaning. Some of these papers were pretentious enough; but most of them are like Etruscan tombs, the preservers of rare and beautiful things which otherwise would not have survived the wear and waste of time. Some one will hereafter collate and classify them for the use of future students. A few of them may be here mentioned, referring to the publications in which they are found by the initial letters.

At the Boston meeting of the American Association of Geologists and Naturalists Dr. Dana read the title of what was, perhaps, the earliest geological report ever made on American Geology, entitled, Contributions to the Mineralogical Knowledge of the Eastern part of North America and its mountain ranges, by Dr. I. D. Schöpf (Beyträge, &c.).

1780, Belknap on vitriol and sulphur, in New Hampshire. (Trans. Amer. Acad.)

1782, Gannett on a yellow mineral paint. (T. A. A.)

1782, Webster on oil stone. (T. A. A.)

1783, Lincoln on the Geology of York River, in Virginia. (Trans. Amer. Acad.)

1783, Gannett and Jones on the West River Mountain. (Trans. Amer. Acad.)

1784, Belknap on the White Mountains of New Hampshire. (Trans. Amer. Philos. Soc.)

1785, Williams on earthquakes. (T. A. A.)

1786, Baylies on Say Head, Martha's Vineyard. (T. A. A.)

1786, Thos. Hutchins on A Cascade near the Ohiopile Falls of the Youghiogeny, twelve miles from Uniontown, Fayette County, Pennsylvania. (T. A. P. S. Vol. II, O. S. p. 50.)

This is perhaps the first recorded geological sketch of any part of Pennsylvania. It occupies but a single page, and was read before the Society January 28th, 1786. It calls the coal measure rock which makes the falls "a species of marble, beautifully chequered with veins running in different directions, presenting, on a close inspection, a faint resemblance to a variety of mathematical figures of different angles and magnitudes. A thin flat stone from eight to ten inches thick, about twenty feet wide, forms the upper part of the amphitheatre over which the stream precipitates. The whole front of the rock is made up from top to bottom of a regular succession, principally of limestone," &c.

1789, Hitchcock on frogs found in the rocks. (T. A. A.)

1793, Franklin on a theory of the earth. (T. A. P. S.)

1799, Dewitt on the minerals of New York. (T. A. A.)

1799, Thomas P. Smith, of Philadelphia, called attention to the "crystalized basaltes" of the Conewago Hills in York County, Pennsylvania, as deserving study. (T. A. P. S., Vol. IV, O. S., p. 445.) This attention was not paid to them, however, until 1820, when Judge Gibson compared them with the Carlisle trap-dyke, and with the rocks at Mount Joy, in Cumberland and Lancaster Counties.

1806, Silliman on the Trap ridges of the Connecticut valley. (T. C. A.) This was a report of the first work he did on his return from Europe, and the earliest attempt of the kind, but one, in the United States. He thus characterizes it in his address at Boston, in 1842; an address replete with facts respecting the early growth of science in this country. But the object now in view forbids a detailed description of work like this done outside of the limits of our State.

1807, Latrobe on Freestone quarries. (T. A. P. S.)

1808–9, Godon on the mineralogy of the vicinity of Boston. (Trans. Amer. Acad.)

1808, Cleveland on fossil shells. (T. A. A.)

1809, William Maclure's Geological Map of the United States, with a memoir, appeared in the Transactions of the American Philosophical Society, Vol. VI, p. 411.

This was the summation of observations made during an extensive tour in 1807 and 1808. It crowned its author with the reputation of being the "William Smith of America," and the father of American Geology. He was an Englishman who had personally examined almost every remarkable geological field in Europe, and was therefore as well prepared as a man could be, at that dawn of geological science, to attempt a sketch of the geology of the New World between the seaboard and the Indian wilderness. A revised edition was published in the Transactions of the American Philosophical Society in 1818, and in a small separately bound volume.

Maclure worked with Silliman around New Haven in 1807, and increased the knowledge which the latter had obtained by a studious residence in London and Edinburgh, between the spring of 1805 and June of 1806, and by numerous excursions in England, Belgium and Scotland. He heard Jameson lecture fresh from the feet of Werner; and Sir James Hall, the successor and expositor of Hutton. Edinburgh was then an arena of excited conflict between the rival schools of fire and water, and Arthur's Seat, Salisbury Crag and Castle Rock stood before Silliman's eyes, representatives of East and West Rocks at New Haven, in view of which his illustrious life was to be passed. One would have thought that the question of the genesis of Trap would have been the first resolved, instead of being left a sphynx's riddle for us still. Let us hope that the younger Dana will be its Œdipus.

Maclure deserves the especial gratitude of the citizens of this Commonwealth for his munificent endowment of the Academy of Natural Sciences of Philadelphia, in its library and its cabinet, which, before the growth of the British Museum and of other later and less imperial collections, was the finest in the world. He lived his latter years in Mexico, and died there, in 1840, aged about eighty years.

1816, Gilmer on the Natural Bridge of Virginia. (T. A. P. S.)

1817, Drake on the Valley of the Ohio. (T. A. P. S.)

1817, Steinhauer on Coal plants. (T. A. P. S., Vol. I., N. S., p. 265.) This memoir is of considerable importance in the history of the Geology of Pennsylvania and of the United States. It is entitled "On Fossil Reliquia of unknown Vegetables in the Coal Strata, by the Rev. Henry Steinhauer;" occupies thirty-two quarto pages, and is illustrated by four copper plates, showing very indistinctly in some cases and very plainly in others, the specific generic characters of Calamites, Sigillaria and Lepidodendra, all of them however under the common name of *stone-plant* (*Phytolithus*, *Martin: P. Verrucosus*, *Dawsoni*, *sulcatus*, *tranversus*, *parmatus*, *cancellatus*, *tesselatus and notatus*). The botanical nomenclature of the memoir is very old fashioned, and the observations of the author equally quaint, but the figures are valuable, especially the fine fig. 1 of plate 6, and fig. 17, "Epidermal impression of *P. parmatus* in ironstone." Seven of the specimens were found in nodules of clay ironstone, or in black-band, and the rest in sandstone. They appear to have been collected from the English coal measures, Low-moor, &c., and brought to Philadelphia, by Mr. Steinhauer. It is not only a curious but an important fact that the author never alludes to American coal measures, in a memoir read before the American Philosophical Society of Philadelphia, the capital of the coal trade of this continent. But then it was read at a meeting in 1817, twelve years before Wittam's memoir was read before the Wernerian Society in London; and eighteen years before the results of Hall's study of the coal plants were published, by Eaton, in Silliman's Journal.

1817. This year is marked by the first appearance of the printed Journal of the Academy of Natural Sciences of Philadelphia. The Society originated on the 25th of January, 1812, when Mr. Thomas Say and a few other gentlemen resolved to meet once a week for receiving and imparting information. The appearance of Wilson's "American Ornithology" and the published papers of Dr. Benjamin Smith Barton had just given a permanent impulse to the study of natural history. Dr. Muhlenberg, Mr. T. Collins, Mr. Nuttall, Dr. Waterhouse, Mr. Say, and Mr. George Ord, were votaries of botany and zoology; Mr. Godon and Mr. Conrad were local geologists; the arrival of Wm. Maclure only lent additional force to their efforts for the advancement of precise science. The Academy was incorporated in 1817, and its Journal then began to be published. In 1825 a new and "spacious" building at the corner of Twelfth and George streets was purchased and its library and museum arranged. Its resident members numbered about sixty in 1831. It built a much larger house for itself, in 18—, at the corner of Broad and Sansom, and in 1873 were laid the foundations of a still more magnificent edifice at the corner of Nineteenth and Race streets. Previous to 1831 *Zoology* was represented in the Philadelphia Academy by Say, Lucian Bonaparte, Leseur, Ord, Harlan, Wood, Green, Coates, Mitchell, Hentz, and Goodman; *Botany*, by Nuttall, De Schweinitz, Elliott, and S. W. Conrad; *Geology and Mineralogy*, by Maclure, Nuttall, Vanuxem, Keating, Troost, Wetherill, Bowen, T. A. Conrad, Seybert and Morton. These men laid a broad and deep foundation for science in the United States west of the Hudson, and in the Journal and Proceedings of the Academy at Philadelphia may be found indications of the progress of all the sciences from year to year to the present day.

1818. Jefferson on the Fossil Bones of Mastodon and Megalonyx, at Big Bone Lick, Virginia. (T. A. P. S.)

1818. J. T. and S. L. Dana on the Mineralogy and Geology of the vicinity of Boston. (T. A. A.)

1820. John B. Gibson, Judge of the Supreme Court of Pennsylvania, on the Trap Rocks of York, Cumberland and Lancaster Counties, Pa. (Trans. A. Phil. S., Vol. II, N. S., p. 156 ff.) Mr. Smith, in 1799, had pronounced in favor of the wet deposit of the trap. Judge Gibson holds an even balance between the watery and fiery theories, but is inclined to think the Carlisle dyke to be a volcanic outflow. In 1838 Dr. Henderson traced this most remarkable of all the rocks of Pennsylvania as a straight and narrow streak, across the North and Cove Mountains, the Juniata and Susquehanna Rivers and Berry's Mountain, cutting through all the formations from I to XI, representing at least 30,000 feet of rock; its width being sometimes not more than four feet. Nor is any notable effect produced by it upon the topography of the country showing side movements along the crack. It waits for explanation now, in 1874, as it did then in 1799.

1821. James, on Trap and Sandstone of the West. (T. A. P. S.)

1824 is the date of the first description of any piece of Pennsylvania geology to be found in Silliman's Journal, (Vol. VIII, p. 236,) entitled A sketch of the geology of the country near Easton, Pa., with a catalogue of the minerals, &c., by J. Finch, No. 126 Broadway, N. Y. It is a short mineralogical sketch, with a small colored map, of the district between the Delaware and Lehigh Rivers and the North Mountain. Chestnut Hill just back of Easton and Marble Hill in New Jersey are evidently the points of attraction to the author.

In this volume of the Journal Chester Dewey's colored map of Western Massachusetts appeared.

In 1825 Dr. Jer. Van Rensselaer, Associate and Lecturer on Geology to the New York Athenæum, delivered before audiences, and printed in an 8vo vol. of 358 pages, six lectures: 1. On Preceding Theories; 2. On Rock Formations; 3. On Coral Reefs, Volcanoes, and other modifying agencies; 4. On the Interior Constitution of Rocks; 5. On Primary Rocks; or, as we should now say, Metamorphism; and 6. On the Transition,

Secondary, and Tertiary Deposits, with synopses of rocks; and Humboldt's, Werner's, and Macculloch's Systems.

This book tells well for the enlightened state of geologists in America in 1825, and must have been of great use in sweeping away false principles of reasoning on observed facts. The author's mild raillery at M. Chabrier's recently published notions, that the great mountain ranges of the earth, as well as the granite boulders of the lowland, are merely fragments of some disrupted planet, and that the water of said planet when spilled over the earth produced the Deluge (page 90), shows that the wildest fancies were shamelessly promulgated even as late as 1825, and that there were teachers who knew how to despise them, themselves, and expose them to public ridicule. The book teems with interesting discussions of theories, and citations of facts; but its apparent but delusive fullness only serves as a foil to that infinite collection of facts, and to that calm discipline of demonstration which has characterized the half century since it appeared in print. If any one wishes to mark the progress made even in the first seven years of this interval, he may read the first article in Silliman's Journal for 1832, vol. 21, in which the editor makes his own statement of the Principles of Geology.

Yet in this latter still appears one of the most obstructive errors of the early geologists. The author, after describing what we now call Glacial Drift, remarks: "The deluge is a great feature in the natural history of the earth. And it is highly desirable to fix the period of its occurrence." Inspired, however, by the instinct of the field-worker, so opposed to the book learning of the cloister, he hurries on to add: "not to estimate how many centuries have passed away since it happened nor how long it remained upon the earth (such knowledge must be gathered from other sources), but its relative place in the succession of phenomena which have visited the earth; for in my mind, those geologists have been ill-advised, who, in the present state of science, affect to form a chronology of nature for comparison with the records of history."

But, after this half-protest against the old superstition, the author goes on to argue out the consequences of a deluge, and to group facts to exhibit them. He wrote at New Haven in

1832. The geologists of Pennsylvania saw the work of the deluge wherever they prosecuted their field work from 1836 to 1841. And to this hour the notion of some supernatural flood-force blinds the minds of teachers and scholars all over the country to the true magnitude of the daily operations of air, frost and running water, and to the real origin of mountain and valley, chasm, cañon, gap and precipice, gravel bed and boulder stone, polished rock and glacial scratch. Nothing has so held back geology as the story of the deluge; and at every step in a retrospect of the history of the science, as relating to Pennsylvania, we may notice the blinding or distorting effects produced, more or less, on all minds looking at things through this old lens.

On page 20 of Prof. Silliman's excellent paper we have a case in point; proving too the consummate genius of old Hutton, and the hardness of the roads which he pointed out for his followers' feet: "No one has carried his speculation so far as Dr. Hutton, who maintained that valleys were, in all cases, scooped out by the streams which run in them. This is a characteristic part of his system of decaying and renewing worlds," &c. "But this opinion clashes so directly with plain facts, as to be wholly inadmissible." Prof. Silliman then instances the dry chalk valleys of England; and roundly asserts that "the excavation of valleys can be ascribed to no other cause than a great flood."

In 1828 Silliman's Journal (XIV, p. 1) published an article "On the Mineralogy of Chester Co.," &c., by George W. Carpenter of Philadelphia, and another "On the Geology and Mineralogy of the Country near West Chester, Pa.," by J. Finch, M. C. C. Dr. Genth's report will signify the value of these papers. The lack of accuracy which characterized the communications received and published by the Journal of that day may be seen by a glance at the romantic portrait of the "Old Man of the White Mountains" which faces page 64; and the ignorance of geology may be measured by the astounding section from Boston water to Lake Erie which faces its title page. No more accurate description of the belt to which the West Chester rocks belong was written until 1833, when Dr. Hayden described (in Silliman's Journal, XXIV, p. 349) the Chrome Iron range north of Baltimore in Maryland, or the Bare Hills, as they were called, A little map is included in this memoir, which shows all the localities of this valuable mineral. The paper is a model of careful description and accurate field work. The author gives 1808, or 1810 as the date of the discovery of "Chromate of Iron" near Baltimore, by Mr. Henfrey. This paper is of the greatest interest to Pennsylvanian geologists; for the Chrome Iron belt reaches Philadelphia, as will appear by Dr. Genth's Report on the Mineralogy of our State, published as part of this Annual Report. About the same time appeared in the same Journal (p. 375), one of the first analyses of the Titanic Iron from Baltimore, by the French Chemist Berthier.

1830. Professor Silliman's notice of the Anthracite beds in the "Valley of the Lackawanna and Wyoming on the Susquehanna" was published in his Journal, Vol. XVIII, p. 308. From this it appears how inadequate were the notions entertained by even the foremost Professor of Geology in America, only six years before the commencement of the first geological survey of Pennsylvania. The carefully engraved map of the valley from Nanticoke to Carbondale is simply an outrage to our present geographical knowledge. The "Ideal section at Wilkesbarre," on p. 309, is as funny as anything in the astrological literature of the middle ages; its naive simplicity affects the weary eyes of the coal geologist to-day like a moonlit street after a hot and noisy "Saturday evening." The Gaylord bed section on p. 325 is given "on account of the peculiar curvature of the bed" (being a gentle anticlinal!). Nevertheless, the "Front view of a contorted coal bed at Pittston" (p. 326) is a valuable relic of what was visible 45 years ago. From the text itself a good deal can be picked up which would elucidate the structure to one charged with more modern knowledge.

In this volume appeared one of the earliest memoirs on the vegetation of coal beds and the standing forests over them by Henry Witham, England, F. G. S. It was read to the Wernerian Society, Dec. 5, 1829.

May 10, 1830, Professor Silliman, with Mr. George Jones, started from New Haven for Mauch Chunk, and he describes what he saw there and at the Beaver Meadow mines in article 1, Vol. XIX of his Journal; giving a curious lithograph picture of the state in which the Summit Mine then was, with an area of eight acres, and two or three stopes around the wall, "exactly as in a stone quarry." It had much the "appearance of a vast fort, of which the central area is the parade ground, and the upper escarpment is a platform for the cannon. The greatest ascertained thickness of the coal is stated at about 54 feet; in one place it is supposed to be 100 feet thick," &c. The whole description is interesting; for the real condition of things was evidently not clearly seen by him, although he says "the geological structure is extremely simple." Mr. White had just opened new mines at Room Run, from data obtained at Summit Mine.

1830. Mr. G. Jones, tutor at Yale, says (in Silliman's Journal XVIII, p. 303), that in a recent visit to Mauch Chunk "he was struck with the universal employment of anthracite in the blacksmiths' shops, and with the strong terms in which the workmen expressed their preference for it over every other kind of coal," and he then proceeds to describe the coal, the furnace and the process. The editor appends a "Notice of the first introduction of Anthracite Coal on the Susquehanna, communicated by Judge Jesse Fell of Wilkesbarre," in which Judge Obadiah Gore, a blacksmith by trade, coming into the Wyoming Valley as a Connecticut settler, is named as the first person who used the coal in a blacksmith's fire, about the year 1768 or 1769.

1830. Dr. W. Meade described a successful experiment on the North River, in burning bricks with the worthless refuse of the anthracite coal yards. (Silliman's Journal Vol. XVIII, p. 118.)

1830. Cannel coal at Steubenville, O., near the west border of Pennsylvania, is first noticed in a letter of Judge Tappan, (Silliman's Journal XVIII, p. 377) dated May 15, 1830. The editor notices the difference of its specific gravity, 1.6, from that of English cannel, 1.2 or 1.3; but, instead of assigning this to a high percentage of ash, imagines that it is more *condensed* than English cannel and therefore more valuable.

1830. Prof. Silliman (Jour. XVIII, p. 210) in reviewing Mr. D. Wadsworth's description of the Upper Falls of the Genesee River, 23 miles above Moscow, north of the Pennsylvania State line, criticises the drawing (which appears in the Journal) in language which deserves recording and pondering over. "It illustrates" he says "not only the picturesque scenery of that interesting region, but also the peculiar geological structure upon which it depends . . . . If a painter were always a geologist, his sketches of rock scenery and of the ever-varying outline of landscape, as it is seen in hills, plains, valleys, waters and mountains, would assume a verisimilitude depending on physical laws, since none of these features are matters of chance; were the geologist a painter, he would breathe into his graphic outlines the living spirit of the sublime and the beautiful." In these few words this most eloquent lecturer on geological science which America ever produced states the law of Topographical Geology. Obedience to this law has placed Russel Smith at the head of American landscape painters; and Prof. Rogers was a fortunate man to obtain for artist of the first geological survey of the State, Mr. Lehman, whose outline-drawings and finished water color and oil paintings were as scientifically accurate as they were artistically beautiful.

1830. Prof. Eaton's "Observations on the Coal Formations of the State of New York," in connection with the great Coal Beds of Pennsylvania, was read before the Albany Institute, March 11th, and published in the Transactions of the Institute, as well as in Silliman's Journal, Vol. XIX, page 21, 24. "It was accompanied with a demonstrative lecture, given at the request of several members of the New York Legislature, while the *bill for boring for coal* was pending." The proposal to bore for coal in New York was not at that stage of American geology preposterous. Dr. Eaton was the author of a Manual of Geology; and yet, in this address, he divided the coal formations of the United States into four: first, the true *anthracite* of Worcester, Massachusetts, and Newport, R. I., in "transition argillite;" second, the *anasphaltic* or false anthracite of Carbondale, Wilkesbarre, Mauch Chunk, &c., in Pennsylvania, in "slate rock, the lowest of the lower secondary series;" third, the *proper bituminous*, of Tioga, Lycoming, in Pennsylvania, &c., in a "slate rock the lowest of the upper secondary series;" and fourth, the *lignite coal* of Amboy, in New Jersey. Such a classification shows how entirely Dr. Eaton's geology was book-learning derived from European authors. When he adds that he and Prof. Van Rensselaer had "carefully traced the slate rock which embraces the bituminous coal of Tioga (Blossburg) to

Seneca and Cayuga Lakes, and down those lakes to their outlets, and to Lake Erie, and along the south shore more than twenty miles," and that "the same bituminous shale embracing the various bituminous coal found in vast beds in Tioga and Lycoming are found in the same continuous rock along the shores of the aforesaid lakes," but that "the thickest of these beds hitherto discovered in the State of New York do not exceed two inches"—a fair estimate can be made of his powers of observation in the field, and a correct appreciation of the necessity for that State Geological Survey of New York, which was commenced six years after he wrote. Everybody now knows that the Worcester, Rhode Island, Blossburg, Towanda, Wilkesbarre, Mauch Chunk and Pittsburg coal measures are of the same age, and have been made to differ in quality only by the subsequent application of heat, pressure, &c. The surveys of Hall and Vanuxem soon established the knöwledge of the great fact that the coal "slate formation" (XIII) of Tioga and Lycoming counties in Pennsylvania overlies by several thousand feet the black (Hamilton) slates (VIII) of the New York lakes.

Mr. Eaton's extraordinary statements were, however, not allowed to pass unchallenged; for Mr. David Thomas, C. E. Cor. Mem. of the Linnæan Soc. of Paris, &c., wrote to Silliman's Journal (XIX, p. 323) from Greatfield, Cayuga county, N. Y., Nov. 15, 1830, very politely, thus: "No person has so minutely examined as large a portion of the United States with reference to its Geology, as Prof. Eaton; and his sagacity has equaled his industry and zeal . . . . . but I wish to ask whether he considers the slate which appears on the shores of our lakes as the same stratum which embraces the Tioga coal? or whether he only means that it belongs to the same (third graywacke) formation? I had been induced to believe that our slate is a different stratum, from considering that *there is a general dip in all our rocky strata to the south.*" And he adds in a note, "The idea appears not to have occurred to Prof. E. at the moment of writing, for he says, '*The layers of this [carboniferus] rock are always horizontal or nearly so.*'" He calls attention to the limestone shores of Cayuga Lake; and again to the building stones at the Seneca locks; and to the *red rocks* (IX), in the valley of Towanda Creek in Pennsylvania, with their salt springs.

Thus the simple good sense of the unlearned confounds the wisdom of the wise; and many a professor of geology is put to the blush by obscure farmers, hunters and miners, who keep their eyes open and their mouths shut, speak little and publish nothing, but do a deal of thinking and do it well; making themselves sure of plenty of facts, and ready to trip up the first comer who is a theorist beyond or in spite of facts. The first Geological Survey of Pennsylvania revealed a wonderful amount of actual science stowed away among the people.

In 1831 appeared Prof. Eaton's description and plate of a fossil scale-tree (lepidodendron) sent to him from Montrose in Bradford Co., Pa. The plate was drawn by Miss T. Lee of the Troy Female Seminary; and the fossil was named by him *Crotalus ? reliquus*, or *Arundo ? crotaloides;* for, as he says in explanation of the double name, if it be an animal it belongs to the genus of snakes called Crotalus (rattlesnake), or if it be a plant it is of the reed family (like *Phytilus Martini*), and deserves to be called the rattlesnake-reed (*Arundo crotaloides*).

In the present advanced state of fossil botany in America, due chiefly to the labors of Leo Lesquereux, Dr. Newberry, and Dr. Dawson, a glance at Miss Lee's drawing suffices to assure even a tyro that it is of a vegetable nature, and to assign it a position among the *lepidodendra.* But the record is, even with Eaton's strange name, valuable; for the country around Montrose contains only rocks of VIII and IX, far below the base of the coal measures. In the paper, however, the author repeats his blunder about tracing the black slates of VIII to the coal mines of Blossburg. He says that the specimen "was found by Dr. Rose of Montrose in Graywacke Rock on his own estate" and that "it lies *over* the Carbondale *anasphaltic* coal," &c.

1831. Notices of the geology of the neighborhood of Bedford Springs, by Dr. H. H. Hayden, were published in Silliman's Journal (XIX p. 27). The author misapplies the term "millstone-grit" to the Oneida sandstone (No. IV) of the surrounding mountains. He notices the Helderburg limestone (VI) at the Springs, but does not name the fossils which he found in them. He notices the Oriskany (VII) over it, "a soft pulverulent sandstone, containing impressions of a variety of shells, as the producti, terebratulæ, a species of pecten, &c. These are the third deposits of organic remains that appear, at least, in order of position." He theories thus: "And consequently, as we ascend in the order of formation and position, the fossils present not only a greater variety, but become more and more complex and perfect in their structure and organization." The effect of book-learning on this observer is very obvious, yet the whole letter breathes the true spirit of an observing and inquiring mind.

November 30, 1831, Pittsburg, Pa., is the date of a letter from Sam. Wyllys Pomeroy, Esq. to the editor of Silliman's Journal (XXI, p. 342), describing the "Coal Region between Cumberland and Pittsburg, and the Topography, Scenery, &c., of that part of the Alleghany Mountains." He writes: "Before daylight we were ascending the Great Savage [Mountain] and had passed the coal bed on the east side from whence it is hauled ten miles to the Potomac at Cumberland." *The* coal bed. Only one bed; only one mine. What a contrast to the present coal trade of the Potomac, 2,000,000 of tons descending to Baltimore annually! Yet, practically viewed, it is still but one coal bed; for the Great Pittsburg bed furnishes almost all of it. "The fracture and general characters very much resemble those of the coal from several localities on James River in Virginia." Here the same absence of all definite knowledge as to the general geological relationship of the later and older formations in America is manifest "I was not able with a strong lens to detect any organic vegetable remains or impressions." Fossil botany had evidently taken its first grasp of the thinking and observing American mind; but its first principles were still unknown. "The sandstone strata which are almost constantly in sight from the base of Shaver's Mountain to near the greatest altitude of the Savage, incline 15° or 20° —the *dip* opposite the setting sun in the middle of November. From thence it declines by gradations hardly perceptible to near the western base, and there becomes horizontal, like all the strata far beyond the Ohio." He did not observe the anticlinal which separates the Cumberland and Salisbury coal basins. "The coal on the Youghiogheny justly deserves the reputation of superiority to any in this whole region. I examined a large heap at Smithfield where we crossed the Youghiogheny, and could not find a single piece that was not beautifully *iridescent* throughout, and exceeded in richness of tints those elegant specimens of anthracite which I viewed in your cabinet." Peacock colors a gauge of excellence in coal! "Owing to the length and the difficult navigation, and low price of coal at Pittsburg it is seldom sent down there and only at high stages of water." Crossing the Laurel Hill he was informed of extensive beds of coal apparently of good quality, and he remarks: "The time probably is not far distant when it will make frequent visits to enlighten its *elder brothers*, the great anthracite formations in the valleys of the Susquehanna and Schuylkill." He had evidently studied the order of coal ages from Dr. Eaton, as he took pains to underscore the words *elder brothers*. He ends by quoting with a little alteration a remark of the editor somewhere in Vol. XIX, "that the sun and the *bituminous* coal of Western Pennsylvania will *burn out together*."

Only two years intervened between the publications last described and one by Prof. Eaton which serves to mark a notable advance in the views of the leaders of the science in America.

In 1833, Prof. Eaton's letter to the editor about "the coal beds of Pennsylvania [as] equivalent to the great secondary coal measures of Europe, appeared in Silliman's Journal, Vol. XXIII, p. 399. He writes: "At the ninety-first page of the second edition of my Geological Text Book, published last June, I adduced facts in proof of the correctness of the heading of this article. Since its publication, Mr. James Hall, Adjunct Professor in this institution [Rensselaer Institute, Troy], has made probably the most extensive collection of vegetable fossils in Pennsylvania that has hitherto been made on this continent. It was the intention of Mr. Hall and myself to have determined the names of all which had been described by M. Brongiart, and to have given lithographic figures of the remainder, but we are prevented by other engagements. At present I will merely give a list of the names of those which we determined by the aid of Brongiart's figures and descriptions as far as his sixth number. I have now before me twenty-five ascertained species of ferns from the coal mines of Pennsylvania, which Brongiart has described as belonging to the great secondary coal formations of

2

Europe, found in the secondary class of rocks *only*. Hence the absurdity of denominating the Allegheny and Catskill Mountains *transition*. If organized remains are any evidence of the equivalent characters of rocks, *these mountains are surely secondary*. They are the upper secondary of some distinguished European geologists, the upper stratum of the lower secondary of others, while others seem unwilling to admit a division of the secondary class. It appears to me that the Allegheny and Catskill Mountains may be assumed, *confidently*, as the grand starting range for settling all questions relating to the equivalent strata of the Eastern and Western Continents. I feel that I am fully supported in the position I have taken in view of such equivalents, and set forth in the last edition of my Text Book, by the additional collections of organized remains made in the months of August, September and October of the present year by the students and assistant teachers of this school. . . . . Every step I take . . . . . and every specimen . . . . . . strengthen my confidence in the opinion of Cuvier, and of the other great men of the East, that 'organized remains are true indexes to geological strata.'" Then follows his list under the heads of Neuropteris, Sphenopteris, Odontopteris, Tæniopteris, and Cyclopteris.

Thus Brongiart taught Hall, and Hall taught Eaton, how to bring American geology into order. Botany preceded zoology in giving stratigraphical law to American geology. Pennsylvania was still the starting point for the accurate geology of the United States. But it was not Pennsylvanian geologists who first saw Pennsylvanian geology in a clear light. For years after the school of Troy had thus formulated the most important generalization on which all our geology is built, the school of Philadelphia prated about *transition* rocks in Middle Pennsylvania. Taylor drew a strong distinction in age between Broad Top and Allegheny Mountain coal; and even Rogers expressed his doubts of their identity in an annual report. Honor to whom honor is due. The same James Hall who has since made the most magnificent American contribution to fossil zoology, made in 1833 the most important of all American contributions to its fossil botany, for it antedated and predicted the immense results of Lesquereux, Newberry and Dawson.

Had structural geology been as far advanced in 1833 as was fossil geology, or had Eaton been a good field observer, he would have discovered the law of succession in his journey from Blossburg to Seneca Lake two years before. It would not have been left to the accident of the publication of the first six numbers of the immortal book of Brongiart to teach American geologists what stared them in the face. For, after all, the fossil plants at Troy merely suggested the idea; it was the seven years of labor in New Jersey and Pennsylvania, from 1835 to 1841—years spent not at all in fossil geology, but wholly in structural geology—that made this order of the rocks an incontrovertible reality. The palæontologists of New York studied the *outcrops* of the formations before they could classify their *fossils*. James Hall stands as confessedly foremost among our structural geologists, as he stands at the head of our palæontologists; and he has always been among the earnest protestors against the pretensions of palæontology to decide *ex cathedra* doubtful questions of stratification.

In 1833 appeared another paper by Prof. Silliman which showed that he also felt the set of the intellectual current, and was free to move with it.

The petroleum fountain, or oil spring, near the county line of Cataraugus and Allegheny, in New York, and near Hick's Tavern, two miles west of Cuba, in Hinsdale township, Allegheny County, was the object of Prof. Silliman's expedition. His description of what is now so well known is of course interesting; but its historical value to Pennsylvanian geologists arises from his confident assertion of a theory of the origin of the petroleum to which, in a modified form, distinguished observers still lend the weight of their authority.

"As to the geographical origin of the spring, it can scarcely admit of a doubt that it rises from beds of bituminous coal below; at what depths we know not, but probably far down; the formation is doubtless connected with the bituminous coal of the neighboring counties of Pennsylvania, and of the West, rather than with the anthracite beds of the central parts of Pennsylvania."

In these words we have a picture of the influence of Eaton's mistake of the continuity of the Blossburg coal through the lake region of New York; and of Taylor's mistake, in making the Broad Top coal and the Anthracite of an older age, and belonging to a deeper formation than the Bituminous. Yet Silliman had already come to reject the theory so far as the anthracite was concerned; and he felt the influence of the neighborhood of the McKean Co. bituminous beds in Pennsylvania. It was a strange, confused jumble of ideas which the men of 1833 tried to put into scientific language and apply to practical uses.

Now that everybody has learned that most of the petroleum springs and wells rise from Devonian rocks (No. VIII) far below all the coal measures (except those of Perry Co., Pa.), some insist upon finding its origin in the black slates of VIII; but as these are in fact the equivalents of the Perry Co. coal measures, Prof. Silliman's language above quoted would suit the views of such geologists very well.

Another interesting passage in this paper gives an idea of the oil trade in 1833 which will be a surprise to many readers. He says:

"I cannot learn that any considerable part of the large quantities of petroleum used in the Eastern States under the name of Seneca oil comes from the spring now described. I am assured that its source is about one hundred miles from Pittsburg, on the Oil Creek, which empties into the Allegheny River, in the township and county of Venango. It exists there in great abundance, and rises in purity to the surface of the water; by dams, enclosing certain parts of the river or creek, it is prevented from flowing away, and it is absorbed by blankets, from which it is wrung," &c.

Returning to the question of subterranean coal beds, he warns his readers that the presence of oil does not *prove* the presence of coal beds under them; but adds that, as horizontal coal beds are not far off in Pennsylvania, they *may* pass underneath the southern part of New York, and some one should bore for them to decide the question.

One more reference will suffice to show the readers of this Historical Report the train of geological thought in vogue in the United States up to 1833.

Mr. James Madison Bunker wrote from Cambridge, Mass., Feb. 23, 1833, to the editor of Silliman's Journal (Vol. XXIV,

p. 172): "Geologists are now convinced that the common bituminous coal, so abundant in the Eastern continent and in some parts of North America, owes its origin to vegetable depositions .... But I believe that many eminent geologists are not satisfied to refer the anthracite formation to the same origin .... Many errors in geological science are justly attributable to an erroneous or limited estimate of time .... It cannot be denied," he goes on to say, "that the power which could create mineral carbon [as some suppose it], could also create vegetable carbon [as it really is], and afterwards, by some great convulsion, subject it to an irresistible consolidating force." The author then says that he had recently been fortunate enough to "obtain from a small quantity of Schuylkill coal six specimens proving that trees were at least present when the coal was formed, if vegetable matter is not its material." These he describes as pieces of wood, "real wood, resembling charcoal, although softer" .... "Either this wood was introduced in some incomprehensible mode into the heart of the solid mass of the coal, or else it is a remnant not wholly consolidated of the material from which the coal was formed." The editor confirms the statement and says, "the structure much resembles that of curled maple."

It will be seen by the foregoing pages, that previous to 1833, there was great intellectual activity manifested in the Northern States by a few cultivators of Geology and the allied sciences of Natural History; that this activity was stimulated and guided from abroad, especially from England; that it was vague in its efforts and erroneous in its conclusions, but capable of self-support and independent discovery; and that the obstacles which retarded its education could be no other than its youthfulness, the lack of reliable standards, and of classified collections, and, above all, the vastness and variety of the New World which its puny strength was resolute to subdue. Give it time, give it numbers, it would find its own opportunities, repeat its observations, correct its first blunders, build up its museums, improve its text-books, consolidate its system, associate its field workers, multiply its learned societies, and soon react upon Europe with an energy and effect as multiform and unerring as the European sciences, developing their own life by similar laws, would be sure to exercise on it.

It must be remembered, among other things marking the era, that the Geological Society of France, with Alex. Brongniart for its President, Desnoyers for its Secretary, and Elie de Beaumont for the first of its Council of Twelve, was born in 1830, and chartered by the King in 1832.

It must be remembered that the Geological Society of Pennsylvania was organized in 1832, and gave an untold impetus to the study of rock formations, their mineral character and worth, their attitudes and serial order, and the organic bodies, animal and vegetable, which were to become our very clock of time.

In 1832, the first notice appeared of the occurence of *marine* shells in the coal measures, putting a new face on the problem of the formation of coal beds. The late lamented John Phillips, of Oxford, one of the foremost English geologists, discovered "Pecten, Ammonites, Orthocera and Ostrea" in the roof-shales of one of the beds of the lower coal measures in Yorkshire. In the Swan Bank's mine, near Halifax a layer of fresh water shells (Unio) was found underneath the layer of marine shells. Lond. and Ed. Phil. Mag., Nov., 1832, p. 349.

In May, 1831, the famous fossils of the Kentucky Big Bone Lick Cave, was brought to the Museum of the Lyceum of Natural History in New York. See Dr. Troost's Memoir in the Trans. Geol. Soc. Pa., Vol. 1, p. 144.

Dr. Green's monograph of Trilobites appeared in 1833. Dr. Harlan mentions the discovery of Calymene Blumenbachii, near Reading, Pa. Trans. Geol. Soc., Pa., Vol. 1, p. 99.

To meet the universal demand for some harmony of the rock systems on the two sides of the Atlantic, based upon fossils, Eaton published, in 1832, his paper on "Geological Equivalents." (Sill. Jour., Vol. XXI, p. 132.) Sowerby, Goldfuss, Cuvier and the Brongniarts were his guides; although in the first sentence, he acknowledges Werner to be his master. But where "relative position" and "mineral constituents" both fail, "fossils" must supply the clue. He refers with pardonable pride to his earliest introduction of the study of fossils into American schools in 1818; and to the lack of books. "But such," he adds, "was the zeal of my students, that I was driven to the work of giving names to our specimens, excepting those which I could make out by the Linnæan descriptions (which I translated and published eleven years ago), and a few labeled by Le Sueur." Then follow three pages of names of fossils, with the formations in which they were found on both sides of the Atlantic. It is a touching little list; like the first boy-sketch of an Orcagna or a Perugino, compared with the crowded gallery of the Uffizi Palace or the Louvre; a list of only eighty names in all, including seven genera of vegetable fossils; but sufficient, he says, when compared with Woodward's synopsis, to prove that the American strata were the true equivalents of the strata of the same names in Europe.

At that time, however, so little was known of the series of animal life-developments from the oldest to the latest geological ages, that every assertion, however false, was credible. Thus in Dr. Harlan's "Critical notices of various organic remains, hitherto discovered in North America," read May 21st, 1834, before the Geological Society of Pennsylvania (Trans. Vol. 1, p. 92), under his notices of a fossil jaw from the New Jersey newer secondary (New Red) rocks, and after saying that many years ago he had received from Mr. A. Jessup a fine collection of fossil fishes found in the slate from Westfield, Connecticut, he quotes an "intelligent friend, a proprietor of a marble quarry situate in Oval Limestone Valley or Nippenose Valley, on the west branch of the Susquehanna River, Pennsylvania, to this effect: "The marble is a greenish colored conglomerate, somewhat resembling verd-antique, and admits of a high polish, being fine grained and hard, interspersed with softer spots of an argillaceous nature. *Some parts of this marble are represented as being replete with the remains of fossil fish*, about the size of a herring or carpe; some specimens retaining the impressions of scales; others only of the bones. The stone was too brittle to permit the obtaining of any of the specimens whole."

Now who, in 1875, would venture to believe the report of a discovery of herring, or carpe, or any other kind of fish in For. No. II, the Siluro-Cambrian limestones of Nippenose or Oval Valleys? No fish in any part of the world have ever been found of an age approaching such remoteness. The author does not even know that these are two different valleys, nor their situation; since he places them on the west branch of the

Susquehanna River; which is only true so far: the two valleys have streams which drain into the river through gaps in the bounding mountain of No. IV. It is of importance, however, that search should be made for this fish locality, because the fish-beds of Ohio are in the Devonian rocks, which carry their outcrops past Williamsport, along the north bank of the river, in front of the Nippenose and Oval valleys.

It is not wonderful that so little was known of the fossil riches of American rocks, when one considers how few eyes were trained to look with intelligent interest on such things, how few minds had then conceived the possibility of their value for determining the age and order of rocks. The great thigh-bone of the Liberty Meeting House (Ky.) mastodon, described by Dr. Troost in the Trans. Geol. Soc. Pa., Vol. I, p. 139, in 1833, had long lain "projecting above ground, and was used in rainy seasons when the run contained water, for a step to cross it, there being a road there also for carts and wagons, which must have fractured many of the [other] bones."

Prof. Cope told us a still more extraordinary story, as late as 1870, of a bridge in North Carolina habitually used by the inhabitants to cross a small stream of water; the bridge consisting of the back-bone of a fossil whale-like animal, fifty or sixty feet long. Thousands of such treasures must be scattered about the United States unknown to men of science, at the present day. One of the noblest specimens of a seaweed-like plant, fossilized in the form of a single stone weighing a ton, probably still lies by a road side in Cambria County, Pa., on the summit of the Allegheny Mountain, back of Tipton.

In 1831 the phenomena of ice action on the rocks was first observed, or at all events made known in a scientific way to geologists in the United States; although no inkling of their real nature and true cause was got until Agassiz published his glacial discoveries in Switzerland.

A letter from Judge William A. Thompson to Prof. Eaton appeared in the July number of Silliman's Journal for that year, describing scratches on the graywacke rocks of Sullivan County, N. Y., wherever the earth had been removed, and ascribing their origin to boulders carried forward by a deluge. He says that about 1820 he fell in with geological works assigning a direction N. W. to S. E. to said deluge; but that his present observations, in more than fifty places, showed that its direction was east and west.

Diluvial scratches were noticed and again described by Mr. John Ball, of Lansingburgh, N. Y., on a mountain in Hebron, N. H., in 1832. (Sill. Jour. XII, p. 116.)

But a most important step forward was made by the present distinguished State Geologist and Archæologist of Wisconsin, in this same fruitful year of 1832. Messrs. D. and I. A. Lapham investigated the nature of the boulders of Ohio, and determined their Canadian origin.

Nor was the time less remarkable for the foundation which was then laid by civil engineers for the mapping of the country. Maps are indispensable to geology; and many of the blunders noticed in the foregoing pages would never have been made in a well-mapped land.

Edwin F. Johnson's Article in Silliman's Journal, Vol. XIX, p. 131, 1831, " On the present mode of conducting land surveys in the United States, was one of the significant expressions of that general and growing sentiment for actual and precise knowledge which brought about Geological Surveys in so many of the States a few years later.

In 1833, the Regents of the New York University published an easy formula for getting the variation of the magnetic needle by observing the moment when the Pole Star and the Star Alioth are in the vertical plane. (Sill. Jour. XXV, p. 262.) The State Geological Survey was then in view, and the necessity for a correct map of the State was pressed upon the attention of the men of science at Albany and Troy.

Crude geological maps like that of Orange County, N. Y. by Young and Heron, published in 1831 (Sill. Jour. XXI, p. 321), and sections like that across Connecticut by Lieut. W. W. Mather (XXI, p. 94), would no longer satisfy the improved scientific taste. Jackson's and Alger's Geological Sketch of Nova Scotia, first published in Silliman's Journal, 1826, when republished in an improved and enlarged form in the Transactions of the American Academy at Boston, in 1832, had to be illustrated by a good map.

Every part of the United States now became a field of inquiry.

It was evidently the dawn of the new day of Geological State Surveys. In 1832, Mr. Alfred Smith published, in Silliman's Journal (Vol. XXII, p. 205), his memoir on the Alluvial and Rock Formations of the Connecticut River Valley, of which he had made a careful study for several years. A map of its water-basin, New Red Sandstone area and trap mountains accompanied the paper. The nearly universal and gentle dip of the New Red towards the east, and its occasional variation to the west, are noted, and its probable fresh-water lake origin (wrongly) presumed; for the fossil foot-marks on the pavement slabs of the Old Church in Northampton had not yet arrested the attention of Dr. Dean or President Hitchcock. But the value of the paper lay in its elaborate section of the rocks revealed by the new canal at Enfield Falls. Beyond this the paper was almost worthless, except as a contribution to the intellectual hunger of the age, and was made wholly so by the publication of Dr. Hitchcock's Report on the Geology of the State, some years afterwards. It helped, however, a little to draw attention to the New Red belt of Pennsylvania and New Jersey.

In 1833, Shepherd published some observations he had made in Georgia, Alabama and Florida.

In 1833, the mining regions of Georgia and North Carolina were sketched in a few pages of Silliman's Journal (Vol. XXIII, p. 1), by Judge Jacob Peck, and on the map accompanying the paper innumerable places are marked "gold." "Titanium" is written on the Tennessee side of the Unaka range; and a "bold vein of hornblende or greenstone" is portrayed crossing the State of Georgia.

The gold mines of Georgia were the subject also of another paper in Silliman's Journal of the same year 1833 (XXIV, p. 1), by Wm. Phillips, C. E., and the cross-sections given in it, although rude, are really valuable, and some of them of high interest to the geologist of to-day. Nor should he be deterred from a careful perusal by fig. 14, the disc of the earth, with "Symmes's Hole" in the middle of it.

Dr. James Dickson read an essay on the Georgia gold region, in June, 1834, before the Geological Society of Pennsylvania (Trans. Vol. I, p. 16), in which the decomposition of the outcrop country to a great depth is prominently brought out; but no conclusions are deduced therefrom, such as Dr. T. S. Hunt has recently published and made applicable to the elucidation of some of the larger problems of the science.

In 1833 also the report of the Committee of the Geological Society of Pennsylvania on the Virginia Gold Region was published in Vol. I of their Transactions, no doubt from the pen of Mr. R. C. Taylor.

Dr. Rush Nott sent to Silliman's Journal, in 1833 (Vol. XXIII, p. 49), a sketchy article on "Miscellaneous geological topics relating to the lower part of the Vale of the Mississippi."

Even from the other side of the Rocky Mountains came geological notes, of very little value, and indeed very absurd, but showing plainly enough how the eyes of all sorts of people were now open to geological phenomena, everybody seeking to explain them on scientific principles. Mr. John Ball sent, in 1833, from the fur traders' posts, a few general geological observations, which were vouched for as scrupulously exact by his old preceptor, Dr. Eaton, and published in Silliman's Journal, Vol. XXV, p. 351. Their utter worthlessness may be imagined from the e sentences: "The geology of the country west of the Rocky Mountains is remarkably simple and uniform. The general underlying rock is the red sandstone . . . . . group of De La Beche. It is the same which contains the salt springs of the western part of the State of New York."

But the year 1832 was noteworthy for yet another reason, to be appreciated all the better. now that geology has ceased to be mere intellectual play, a transcendental excursion into the charming world of planetary genesis, and has become, instead of that, the slave of economy; a guide to the treasures of force; a fosterer of the comfort of the masses of mankind; the fee'd expert of the Iron Manufacture; and a respected friend of money-makers on 'Change.

Walter R. Johnson, Professor of Mechanics and Natural Philosophy in the Franklin Institute at Philadelphia, began, in 1832, that course of experiments on the steam boiler which culminated in his classification of the various coals according to their heating powers, in a book which made him one of the most famous men of science in America.

An article in Silliman's Journal, XXI, p. 71, 1832, "On the variable rapidity of action between water and hot iron," was followed, in 1833, by another: "On Economy of Fuel with reference to its domestic applications" (XXIII, p. 318), and in 1843 appeared that work on Coal which has formed the basis of all subsequent researches of the same nature.

CHAPTER II.

THE GEOLOGICAL SOCIETY OF PENNSYLVANIA; AND WHAT IT DID TO BRING ABOUT THE FIRST GEOLOGICAL SURVEY OF THE STATE.

In the early spring of 1832, seven men of science met in Philadelphia and organized the Geological Society of Pennsylvania.

Mr. Peter A. Browne was a leader in this movement, so important in its results for the scientific and practical study of the mineral resources of the Commonwealth, and was elected to be its Corresponding Secretary.

John B. Gibson, Richard Harlan and Henry S. Tanner appear as a committee to memorialize the Legislature for a Geological Survey.

The officers elected at its first meeting, dated Feb. 25th (22d in another account), 1832, were these:

John B. Gibson, President.
Nicholas Biddle, Vice-President for Philadelphia City.
Samuel H. Long, Vice-President for Philadelphia County.
Henry S. Tanner, Treasurer.
George Fox, Recording Secretary.
Peter A. Browne, Corresponding Secretary.

The Society fast grew in numbers and influence. The list of its members given at the end of the first part of the first volume of its Transactions (the only volume it ever published), bearing date 1834, states their number as follows: Resident members 83; correspondents in Europe and America 32; honorary members 4. According to a report of the Committee of Inspection, dated Feb. 25, 1835, quoted at the end of the second part of the same volume (published separately in 1835), it then "numbered more than 200 resident or corresponding members, and 4 in the honorary class." Its "efforts were seconded also by several local institutions which had been established in some of the counties of the State, for the promotion of a general State survey."

This was evidently the principal object of the creation of the mother society, whose life was so brief, but out of whose cast chrysalis skin issued, at Philadelphia, in 1840, the "Association of American Geologists (and Naturalists);" which again, in 1848, and for the third time at Philadelphia, became by name "The American Association for the Advancement of Science," still existing in full vigor, the most important learned scientific organization in America, holding its annual summer meetings in the different cities of the United States.

The organization of the Geological Society of Pennsylvania, in April 1832, marks then the beginning of a new era in American science, and agrees in time with the commencement of that series of State Geological Surveys, which up to 1843, had been carried on in three-fourths of the States of the Union.

The Constitution of the Geological Society of Pennsylvania is given on page 208 of Hazard's Register, Vol. IX, for 1832; its circular letter to the citizens, on page 306, of Vol. X, for 1832; and its memorial to the Legislature, with the report of the Legislative Committee on a Geological Survey, on page 225 of Vol. XI, for 1833.

The Constitution reads as follows:

"The objects of this Society are declared to be, to ascertain as far as possible, the nature and structure of the rock formations of this State—their connection or comparison with other formations in the United States, and of the rest of the world; the fossils they contain—their nature, positions and associations, and particularly the uses to which they can be applied in the arts, and their subserviency to the comforts and conveniences of man.

"To effect these desirable objects, its members promise to contribute their individual exertions, and to use their influence to have the State geologically surveyed, to assist in making a State geological mineralogical collection, to be geographically arranged, at such place as the Society shall appoint; and to disseminate the useful information thus obtained by geological maps, charts and essays.

"The Society shall consist of such persons as may subscribe

to this Constitution, and such others as shall hereafter be elected agreeably to this Constitution and the By-Laws hereafter made.

"The Academy of Natural Sciences of Philadelphia, the Chester County Cabinet, the Cabinet of Science of Bucks County, the Cabinet of Natural Science of Montgomery County, the Library and Reading Rooms of the Northern Liberties, the Cabinet of Natural Science of York County, the Cabinet of Natural Science of Bradford County, and such other similar societies as shall be hereafter erected in this State, under the auspices of this Society, shall have a right to nominate to us, annually, one of their members, who (unless some good reason can be given to the contrary) shall for the time being, enjoy all the privileges of members of this Institution.

"This Society shall hold stated meetings, twelve times a year; and adjourned meetings as much oftener as they shall think proper. Four meetings at least, if practicable, shall take place at the following places, viz.: Philadelphia City, Pittsburg City, and the Borough of Harrisburg."

Other paragraphs fixed the number of officers, and the fees of membership.

The Circular Letter, which was issued by the Society, inviting the co-operation of the citizens of Pennsylvania, stating the objects which the Society kept in view, ran thus:

"To have an exact knowledge of the mineral resources of this State, is considered the most important of these objects, and as it is the intention of this Society, to construct, as soon as the proper information is obtained, an accurate Geological Map, which shall indicate the mineral topography of the State, you are respectfully requested to return at your earliest leisure, answers to the following queries, and to assist in giving effect to the intentions of the Society, by procuring and furnishing them with the information and specimens now solicited, as far as your opportunities and convenience may admit of. As the proceedings of this Society will be occasionally published in the *Monthly Journal of Geology and Natural Science* of this City, the valuable information thus procured will be publicly acknowledged, and its authenticity be satisfactorily established."

The queries which follow number twenty-eight. They ask

the residence of the correspondent, and the names of well-disposed individuals, and of known and competent surveyors. They call for correct geographical sketches of mountains, valleys, &c., in each county; whether there be coal, iron, lead, copper, marble, limestone, or other valuable minerals; and their localities, qualities, minable areas and limits. They invite collections of "some of the most perfect fossil coal plants, a specimen of each variety, and the localities noted." "Upon what general bed does the coal lie? Is it limestone, sandstone, clay, shale, or what other simple mineral?" (This was to settle the quarrel about the age of the Coal Measures.) "Will you procure for the Society geological specimens, not exceeding four inches square, of the general bed under the coal field mentioned in query 11, as well as of the alternating beds spoken of in query 12, together with good specimens of every species of organic remains found in all such beds, noting their localities?" (A round request, indeed, in a day of no railway, express or transportation companies.) Then, questions about salt springs, mineral waters, rock salt, depth of wells, sections of bore-holes. Then, a second request for specimens of rocks and fossils from the mountain rocks; for perpendicular sections of cliffs and river banks, with specimen rocks and fossils; for information about caves and bone-beds and stalagmite coverings; (the story of Buckland's Kirkdale Cavern was bearing transatlantic fruit); and about any skeletons not found in caves, and how buried—in clay, marl, sand or gravel, with or without shells or broken trees. The Society wanted such skeletons very much, or if not the skeletons themselves, then good drawings of the same. "Please wrap all specimens carefully up," was the closing petition, "and forward them packed in a box, by the cheapest and earliest opportunity, addressing the package to Peter A. Browne, Esq., Corresponding Secretary, and giving information by mail of the time and manner in which the package was sent. Signed by order of the Society, John B. Gibson, President, and George Fox, Recording Secretary; Philadelphia, March 1st, 1832."

The memorial of the Society to the Legislature of 1832–'3, was written by Mr. Peter A. Browne, and is thus summarized by Mr. Say, Chairman of the Committee of Legislature appointed to consider it:

"The memorialist proposes to make a Topographical, Geological and Mineralogical Survey of the State, to publish a complete series of geological maps, profiles and sections; and to form the scientific collections, to be deposited in seminaries of learning, and other places, where they can best subserve the purposes of instruction and practical usefulness, in aid of which the State is asked to subscribe for one thousand copies of the maps or atlas, which will be divided into twenty-seven numbers each, at one dollar a number, amounting to the sum of $27,000; and he also proffers to place at the disposal of the Legislature a complete cabinet of specimens of all the rocks, fossils and minerals that shall be found in Pennsylvania while making the survey, and a scientific report of the same."

Little did this bold petitioner know the dangers of disaster which he faced in making such a proposal at that early day. But men would undertake few enterprises had they the gift of prescience.

The committee, after alluding to petitions from citizens of various Counties of the State, from the Cabinets of Natural Science of Lancaster, and of Montgomery County, and from the Geological Society of Pennsylvania, go on to say, that they considered it inexpedient to entrust so arduous and important an undertaking to any single individual, and had therefore applied to the Geological Society of Pennsylvania, inquiring "whether the Society would find it consonant with their views to undertake the direction and responsibility of the work." The Society had replied, "That they will receive and assume the responsibility for the faithful application of any sum of money granted by the Legislature for the said survey."

"They, the Society, name, in addition to the objects proposed to be effected by the contemplated measure, to establish three meridian or transit lines, extending entirely across the State, to be denominated the eastern, the middle and western meridians of the State, measured with the utmost care and precision, and permanently marked at intervals of a mile, or oftener, on suitable posts or stones, set for that purpose."

Turning to the full text of the Society's reply, which accompanies the report of the committee, this project for establishing meridian base lines to serve as a foundation for a good State Map is argued out in detail.

The eastern line should commence near the northeast corner of Maryland and cross Chester, Berks, Lehigh, Northampton, Luzerne and Susquehanna Counties, passing near Coatesville, Reading, Mauch Chunk and Wilkesbarre to Great Bend.

The middle line should commence near Hancock and crossing Bedford, Huntingdon, Centre, Clearfield, Lycoming and Potter Counties, should pass obliquely across numerous ranges of mountains, and near by Stonerstown, Williamsburg, Huntingdon, Birmingham, Philipsburg, Karthouse, Coudersport and Port Allegheny.

The western meridian should leave at right angles the Virginia State line at the Monongahela River, and crossing Greene, Fayette, Washington, Allegheny, Butler, Venango, Crawford and Erie Counties, to Lake Erie, pass near Greensburg, Newport, Greenfield, Brownsville, Bentleyville, Pittsburg, Butler, Franklin, Meadville and Erie.

There was a spice of diplomacy in all this. However distant such meridian lines might lie from any of the above named places, the mention of their names would be likely to induce their inhabitants to instruct their representatives at the State capitol to further the survey. In fact, the advantages held out were not unreal; especially that of the easy rectification of the magnetic needle for running old surveys. "Internal improvements" were also hinted at; canal and railway surveys, which however would not be much assisted by three meridian lines drawn over hill and vale, even if leveled with all needful care. Nor were the possible geological benefits to accrue from three base lines very great, as subsequent surveys have demonstrated. Isolated straight lines, however well run and leveled, are of little use in geology. Nothing but a close net-work, or gridiron system, of such lines avails to reveal structure; and in Pennsylvania even such a system is good for little when oriented due north and south But all such knowledge came with the experience of after years; in 1833 there was none of it.

The committee goes on to argue the value of the survey to the Commonwealth, and to quote the example of three sister States, Tennessee, South Carolina and Massachusetts, which had commenced their respective State surveys, under the direction of Dr. G. Troost, Mr. Lardner Vanuxem, and Prof. Edward Hitchcock. The whole report is well written, and shows advanced views both of a scientific and business kind.

"*An act providing for a Geological Survey of the State*" was then reported and read as follows:—

"Section 1. Be it enacted," &c., &c., "That if the Geological Society of Pennsylvania shall, within 60 days after the passage of this act, engage, by writing, under their corporate seal, to take upon themselves the duty of causing to be accurately located and designated, at least three meridian lines, crossing this State, and one other line at right angles therewith extending through the State, and shall in like manner engage to conduct a geological and topographical survey of the State and to furnish for the use of the State [blank] copies of the map, profiles and sections of such survey; in such case the Governor be and he is hereby directed to draw his warrant on the State Treasurer in favor of the President, Vice-President and Secretary of the said Society, for such sums of money, not exceeding $15,000 in all, and not exceeding $5,000 in any one year, as he shall judge proper, to be expended by the said Society for the purposes aforesaid: *Provided*, however, "that the manner of making such location and survey shall be first submitted to the Governor, and by him approved."

Could any appropriation be more modest, considering that it was to be made by the "wealthiest mineral State of the Union." The Legislature considered the subject in its session of 1832–3; and considered it again the following year 1833–4; but when at the close of the session of 1834–5, the Committee of Inspection of the Geological Society made their annual report (Feb. 25, 1835), they write:

"This Society has again to regret the further postponement of this all-important measure by the State Legislature. During the recent session, discussions relative to the political state of our country generally appear, in too many instances, to have occupied the attention of the members, to the exclusion of measures of permanent utility.

"In the meantime, the Legislatures of our sister States have shown increased interest in obtaining a correct knowledge of the mineral wealth of their respective States. In addition to those portions of our country whose Legislatures have already availed themselves of the scientific labors of native geologists, we are now enabled to add the States of Virginia, New Jersey, New York, Connecticut and Maine, where active measures are at present in operation to secure complete geological surveys."

One year more was spent in the manufacture of public sentiment by men of science and enlightened men of business, and then, in the spring of 1836, a resolution passed both houses of the Legislature and was signed by the Governor; this was four years after the date of the presentation of the memorial of the Secretary of the Geological Society of Pennsylvania.

Of the Society itself we hear no more. After accomplishing the sole object of its creation it seems to have gone to rest. But if its corporate body died, its spirit continued active. Its members haunted the halls of the Academy, and at length, in 1840, as has been said above, organized the Association of American Geologists. But they took no part in the Geological Survey of the State. The Geological Society, after all its efforts, was never entrusted with the work. In fact, none of its members, singly, was capable of such a task; and no society habituated to debates, collectively, could be. The direction of the survey was entrusted to a stranger; but to one at least as much at home in Pennsylvania as any of the geologists of the Society; to one bred in the same kind of country; to one who, as State Geologist of New Jersey in 1835, had prepared himself for the survey of at least a part of Pennsylvania; to Prof. Henry Darwyn Rogers, the brother of William B. Rogers, the State Geologist of Virginia.

The report of the Inspecting Committee of the Geological Society sums up the efforts made by the Society in a few significant sentences deserving record: "Documents had been prepared, at considerable expense, to explain the intentions of the Society; two qualified members had been sent to Harrisburg; weekly lectures on geology were delivered to the public; minerals, sent to the Secretary by citizens of the State, had been analyzed; five hundred copies of a volume of Transactions, (I. i.) consisting of 180 pages and 6 plates, were extensively distributed; and as many copies of reports of committees and papers of minor interest. In the winter of 1834–'5, another volume of Transactions (I. ii.) of 248 pages was printed and distributed; and duplicate copies of these rare volumes are now to be seen in the State Library at Harrisburg.

The Transactions of the Geological Society of Pennsylvania stand in evidence of the intellectual activity of that group of geologists and naturalists which made Philadelphia, and its Museum of Natural History, in that day, the chief centre of scientific investigation in America. These Transactions were geological reports of various districts, of various value, variously illustrated, read at the meetings by Richard Cowling Taylor, of England; Peter A. Browne; Mr. G. W. Featherstonhaugh, of England; Dr. Richard Harlan; Dr. Jacob Green, Professor of Chemistry in the Jefferson Medical College; Mr. Jas. Dickson, of England; Andres del Rio, Professor of Mineralogy in the School of Mines of Mexico, (President of the Geological Society in 1835); Dr. Gerard Troost, Professor of Chemistry &c. in the University at Nashville, (one of Napoleon's savants in Egypt); Mr. T. G. Clemson, a young student of the Paris School of Mines, (just appointed Superintendent of the Flemington Mines in New Jersey); Mr. Edward Miller, afterwards one of the most distinguished civil engineers of the United States; Mr. Timothy A. Conrad, afterwards palæontologist of the New York Survey; Mr. H. 'Koehler; and Mr. James P. Espy, the famous meteorologist. With these men were associated Prof. Walter R. Johnson, Mr. Sears C. Walker, Mr. Isaac Lea, and others equally noted in the world of science. Among the officers were its vice presidents, Henry S. Tanner and Samuel H. Long, J. M. Brewer, N. C. Sinquet, C. S. Miller and W. P. Gibbons.

Many of the papers read at the meetings were first published in the pages of the *Monthly Magazine of Geology*, edited by Mr. Featherstonhaugh, who, in 1834 and 1835, quitted Philadelphia to make reconnoisance surveys in the far west for the United States government, and at an advanced age, died recently in France. One of his papers in the transactions of the society gave a rambling description and a worthless geological section across the continent from New York to Texas.

In July, 1831, appeared the first number of the first volume of Mr. G. W. Featherstonhaugh's *Monthly American Journal of Geology and Natural History*," published at Mr. H. R. Porter's, No. 121 Chestnut street, Philadelphia. One "continuous essay on geology," was promised for every number; and the vignette to the first number was a lithograph of the cast of a fossil fragment of a jaw of Rhinoceroides Alleghaniensis, found beneath the roots of an oak tree, on Castleman's river, 13 miles above the Turkey Foot, in Somerset county, Pennsylvania.

A map and discussion of the erosion of Niagara Falls; and a repudiation of the amusing French geological nomenclature proposed under the word "Theoric," in the 54th volume of the "Dictionnaire des Sciences Naturelles," appeared in the same number.

Dr. Harlan's *Fucoides Alleghaniensis* from No. XII, is also mentioned.

In the September number appeared a long article "on the value of geological information to engineers," &c., and a savage attack upon Dr. Hays, for defending his friend Godman, the namer of *Tetracaulodon Mastodontoides;* from which it appears that Mr. Featherstonhaugh had delivered a course of twelve lectures on geology in Philadelphia, in the spring of 1831, as he had previously done in New York, "for the sole purpose of advancing the cause of natural science in this country."

His account of the numerous and curious fossils found at Big Bone, Ky., in the October and November numbers, is valuable. He insists much upon their promiscuous and fragmentary assemblage, but does not connect the phenomenon with that of sink-holes, which has played so important a rôle in the erosion of the Western country, and of Pennsylvania also. He however mentions the artificial looking "heaps" of tusks and teeth found, in 1816, near Canstadt on the Neckar, and at Thiede in Brunswick.*

Professor James Hall, in his memoir on the mammoth remains found in the pot-holes at Cohoes Falls, N. Y.,† speaks of gla-

* Buckland Relig. Dil., p. 180.

† Twenty-first Annual Report, Regent's University, N. Y., 1871, p. 108.

cial scratches found on such remains, and makes ice play the principal rôle in all phenomena of this kind.

In the December number appears a description of Harlan's *Fucoides Brongniartii*, as found "in compact sandstone subjacent to the coal formation," and yet "near Lockport, N. Y." This was five years before the beginning of the New York geological survey.

A paper on the primary rocks of Philadelphia, as a continuation of those of Connecticut, appeared in the December number, in which the editor deprecates the misapplication of the term *trap* to the felspar-hornblende rocks of the Philadelphia-Baltimore range; and also the use of the vague word *diabase*, which may, with equal propriety, be applied to any double-base rock. He says: Hornblende-rock is a quite good enough name. He considers the English names of Conybeare and Sedgewick as applicable on one side of the Atlantic as on the other.

In the January number are shrewd thoughts respecting the formation of Buckeye Tunnel, Scott county, Va. And in the March number is a good picture of the Natural Bridge, Va.

In the February number (1832) appeared a paper "On the geological character of the Beds on which Philadelphia stands," by Peter A. Browne, Esq.

In the March number appeared an interesting resumé of the societies and magazines of science published in 1832 in the United States; and in this, mention is made of "the American Geological Society," established at New Haven, Connecticut, "and founded many years ago," but then defunct; its president, Wm. M'Clure, resident in Mexico; and the society itself without a house, collections or transactions.

The circular letter of the Geological Society of Pennsylvania to citizens of the Commonwealth, voted at its meeting, February 22, 1832, (after electing John R. Gibson, president, Nicholas Biddle and Stephen S. Long, U. S. A. vice presidents, Henry S. Tanner, treasurer, Peter A. Browne and George Fox, secretaries,) was published in the March number of Mr. Featherstonhaugh's magazine. It called for information under 28 heads; wishing to be informed if the recipient would furnish a geological sketch of his county, or answer to such questions as: Who was the best surveyor in his county? Would he sketch its topography? Were there coal, iron, lead, &c., &c., known to exist in it? What character of mineral it was? Would he collect fossils and minerals? Would he make sections of the coal measures? Get records of salt wells, cliffs, &c.? Search caves for fossil remains? or procure drawings of objects of science for the society and transmit such to the secretaries.

The society was evidently bent upon obtaining by hook or by crook some sort of geological map of the State, and it promised to make full and honorable acknowledgments in the monthly issues of Featherstonhaugh's Geological Magazine, for all assistance rendered.

[At this time appeared Dr. C. T. Jackson and Mr. Alger's geology of Nova Scotia (116 pages) in better style than anything yet given to the public.]

In the April number of 1832 appeared Mr. R. C. Taylor's "Section of the Allegheny Mountain and Moshannon Valley, in Centre county, Pennsylvania;" illustrated by one of those wildly exaggerated profile drawings in vogue forty years ago, but wholly foreign to Mr. Taylor's own taste; as any one may learn who will glance at the beautifully drawn and colored sections in his "Geology of East Norfolk," published in London in 1827. The illustrations of his early papers in the Transactions of the Geological Society of London, Sec. Ser. Vol. II, are strong, but not so neat. This memoir had been read before the Geological Society of Pennsylvania at the meeting of April 14, 1832.

In the same number a report of the meeting of the British Association gives the editor an occasion for contrasting Hutton's views of the great "whinstone" 100 mile trap dyke of North England, with those of Murchison & Sedgewick; and he consequently prefixes to his March number instructive sections of *injected trap beds* borrowed from M'Culloch's "Western Islands." This plate had its effect in determining the views of Mr. Rogers and other American geologists, when they came to study, a few years afterwards, the trap ranges in the New Red of the Connecticut river valley, Central New Jersey, and Bucks, Berks, Lancaster, York and Adams counties, Pennsylvania.

In the same number is a minute of a communication read at the April 14, (1832) meeting of the Geological Society of Pa.

by a *Committee of the Cabinet of Science of Bradford county, Pa.* respecting the bituminous coal and iron ores of that district, and traces of copper. "Major Long, of that county, had detected gold and silver in particular rocks," &c. The Towanda mines were then wrought to some extent and the coal sent on snow-sleds to Ithica.

A communication on the geology of Wayne county, Pa., with a map and section, from Jacob P. Davis, of Bethany, was also read at that meeting, and is a very curious specimen of the kind of fruit naturally to be expected from such a circular as that which the society had published in all parts of the State. It shows also that there were minds everywhere awake to the desirableness of a State Survey.

In the June number of 1832, appeared Dr. Jacob Green's Synopsis of the Fossil Trilobites of North America, with ten of them rudely lithographed on a frontispiece plate. The central figure (*paradoxus spinosus*) was perhaps intended for the since so celebrated *Paradoxides Harlani*, and believed for a long time to have been dropped on the Boston wharves from the ballast of some European ship, until a Boston banker, who had purchased a country house near the Quincy granite quarries, found multitudes of them in the quarries, in stone fences, and on farmer's mantle pieces, recognized them as trilobites, and showed them first to Dr. C. T. Jackson and then to Prof. W. B. Rogers. Prof. Rogers immediately visited the locality, obtained specimens and described the species at the next weekly meeting of the Boston Natural History Society. It then became known that thousands of these first found fossils of Eastern New England had been for years systematically pitched into Boston harbor to make foundations for the wharves.

Mr. R. C. Taylor read the first paper published in the Transactions of the Geological Society; and afterwards, others of considerable value. Their titles are as follows:

1. "On the geological position of certain beds which contain numerous fossil marine plants of the family fucoides, near Lewistown, Mifflin county, Pennsylvania," being the substance of a description of Dr. Harlan's *Fucoides Alleghaniensis*, which he communicated with a drawing of the fossil to the London Magazine of Natural History. Mr. Taylor had afterwards found it

in the brown sandstone (Clinton, No. V,) of Tussey mountain, near Alexandria, Huntingdon county, and again in Bedford county. He says it prevails in the white sandstone (Oneida, No. IV.) of the Seven Mountains in Centre county, and he had obtained splendid specimens in the white sandstones near Muncy in Lycoming county. He had also observed *fuci* in the old red (Catskill, No. IX) of the Allegheny mountain, at long distances apart. Others had been noticed in Long Narrows, below Lewistown, on the Juniata river, where the Pennsylvania canal had been excavated at the base of Shade mountain, (Oneida, No. IV,) and here "in singular abundance." He traced these beds October 2d and 25th, 1832, two miles, and counted seven courses of fucoid beds within a thickness of only four (4) feet. On November 11th and 12th he found twenty layers in a thickness of three (3) feet. "There seems to be more than 150 feet of the series." His whole description is worthy of study; but the immature views of the geology of Pennsylvania which it exhibits, and which this indefatigable and intelligent field geologist shared in common with his cotemporaries, are equally note-worthy. Whilst he recognizes the synclinal (of V) in which the Juniata flows through the Long Narrows, he asserts that the two bounding mountains not only incline at opposite angles, but "are formed of a different material," and "the beds broken and distorted by violent action and apparently uncomformable to the subjacent rock (!) bear evidence of a more recent origin."

2. "On the relative position of the Transition and Secondary Coal Formations in Pennsylvania and description of some Transition Coal or Bituminous, Anthracite and Iron Ore beds, near Broad Top Mountain, in Bedford county, and of a coal vein in Perry county, Pennsylvania."—*Trans. Geol. Soc. Penn'a, I ii, p.* 1.

Four delicately engraved sections on one plate accompanied this memoir, which show in an instructive manner how utterly a really good observer can overlook the prime facts of structure in a difficult country. The limestone of Kishacoquillas Valley is made to descend in one vast mass so as to separate the Stone mountain rocks from the Jack's mountain rocks by a great age. Of course such sections must have been in good part guessed at from a distance. The section from Tussey to Broad Top is fairly stated; and the contorted strata on the Juniata are beautifully displayed. But this paper possesses a peculiar interest to us now, showing, as it does very clearly, that Mr. Taylor localized the Broad Top coal in far older ("graywacke") rocks than the Allegheny Mountain Coal Measures; and this is quite enough to convince us how entirely the true structure of the State was misconstrued in the years previous to the First Geological Survey.

This memoir takes precedency, in the Second part of Vol. I of the Transactions of the Geological Society, and was no doubt read in 1834. It commences with a reference to the "Report of the Committee of the Senate of Pennsylvania upon the subject of the Coal Trade," and reproaches the author of that "valuable report, which combines a greater mass of useful practical information on the subject of Pennsylvania coal than has ever before been presented to the public," with having repeated, on page 33 and on page 122, the error of "classing the coals of Cumberland, Wills' creek and the Round Top (Broad Top?) mountain . . . with the Secondary bituminous coals of Clearfield and Lycoming." He adds: "having had some opportunities of investigating the relation, ages and the geological order of position in most of the deposits referred to, I feel no hesitation is assigning the Bedford county coals, particularly the veins of the Broad Top mountain, to their true position among the grauwacke, or as they are commonly designated the transition rocks," &c.

We have here the curious spectacle of the first geologist of the Geological Society of Pennsylvania dogmatically condemning the first clear utterance of a great geological truth, and the society lending the weight of its imprimatur to the support of an old and obstinate error.

This memoir of Mr. Taylor exhibits some of the earliest steps taken to unravel the labyrinth of anticlinal and synclinal axes of Middle Pennsylvania; and the author's blind wanderings in that maze can best illustrate the marvellous patience and skill of Dr. Henderson in 1839 and 1840, and the grandeur of his work as exhibited in his colored manuscript map, which has never been published, but was used in the construction of the State Geological Map.

But by far the most interesting narrative in the memoir is that (on page 188) where Mr. Taylor describes the Oldest Coal Measures, lying at the base of the Devonian system, (Hamilton, bottom of No. VIII,) which cross the Juniata river in Perry county, 28 miles above Harrisburg; a coal formation peculiar to Pennsylvania and to Middle Pennsylvania west of the Susquehanna river; a coal formation several hundred feet thick, and traceable along the Lewistown valley. Mr. Taylor says that "within the last year (1833) researches have been made with some perseverance by Dr. Martin, on the west side of the Susquehanna, and nine miles above the confluence of the Susquehanna and Juniata rivers, in a prolongation of Berry's mountain, which forms the southern boundary of Lykens valley," the northern fork of the Pottsville Anthracite Coal Field. "The same vein has been found below Millerstown."

Mr. Taylor then compares the coal of this vein, too thin to work, with that of Lykens valley, and with that of Broad Top, and as his theory demanded, likens it rather to the latter. He adds: "It seems, therefore, that the quality of Broad Top mountain coal depreciates as it advances northward. No decided vein of workable coal has been proved between the Juniata and the Susquehanna, a circumstance which is singular; because, in a geological sense, I know no cause for the absence of carbonaceous deposits within that extensive area; and have reasons, which acquire strength with renewed observations in that quarter, for conceiving that they will ultimately be found there."

Thus a false theory feeds illusory hopes. Taylor could not see that the Broad Top was an outlying patch of the great Bituminous Coal Area of Western Pennsylvania, cut off from it by erosion along the great anticlinal of Morrison's Cove and Canoe valley; and that all the rest of Middle Pennsylvania was a labyrinth of much older rocks, in which an older coal system indeed existed along the lower Juniata valleys, but was utterly worthless for economical purpose; having regular coal beds, roof slates and intervening shales, sandstones and conglomerates, but not one of its half dozen or dozen beds more than a few inches thick. At intervals of a few years the old errors of Mr. Taylor have reproduced the old efforts of Dr. Martin, and large sums of money have been spent, even as late as 1860, long after

Dr. Henderson's thorough elucidation of the geology of the whole region, by men who asked no advice, read no geological books, and listened only to the foolish gabble of miners out of employment, or wandering adventurers willing to drive tunnels in any rock, and open coal beds an inch thick, so long as credulous capitalists at a distance, would give them bread and butter and a roof over their heads.

In a geological point of view, however, all such abortive explorations are as valuable as they are expensive; and although Dr. Martin threw away every dollar of his costly adventure, the Geological Society of Pennsylvania "was indebted to him for a splendid series of fossil coal plants from this mine."

And this splendid series of fossil plants became in turn, oddly enough, another snare. "Dr. Harlan, who has examined them attentively," says our author, "thinks they can be referred to *Knorria imbricata*, *Lychopodiolites dichstomus*, *and Striaticulmus*, figured by Sternberg," and known in Europe as Devonian fossils. This only confirmed Mr. Taylor in the belief of the Devonian, or grauwacke age of the Broad Top coal, although no such plants could bé discovered in them, but because he could imagine no reason why the Broad Top basin should not extend across the Lower Juniata to the Susquehanna river.

By the First Geological Survey under Prof. H. D. Rogers, all these errors were swept away, and no geologist now feels embarrassed at anything he sees in Middle Pennsylvania. But it still remains a desideratum that these fossils, if they can be discovered in any cabinet, and identified as having actually come from Dr. Martin's mine, should be studied and described by Professor Lesquereux, the fossil botanist of the Second Geological Survey of Pennsylvania.

3. "Notice as to the evidences of an ancient lake, which appears to have formerly filled the limestone valley of Kishacoquillas, in Mifflin county, Pennsylvania." A few neat little landscapes accompany this paper, and among their stiff mathematical lines can be detected the curious crests of the "keel mountains" at the east end of the valley. But the author was blind to the origin of the terraces which surround them, because the idea of erosion had not then been conceived by any American geologist. Since the study of lake terraces, river

terraces, glacial moraines, rocky mountain parks and all that world of quarternary geology, no one can now look at the beautiful marginal terraces of our interior limestone valleys with Mr. Taylor's eyes, any more than a merchant or a machinist can see boats and kites as a child sees them. Mr. Taylor's colored section of the valley is a curiosity; rich in historical suggestions of the progress of Pennsylvania geology.

4. "On the Mineral Basin or Coal Field of Blossburg, on the Tioga river, Tioga county, Pa."—Mr. Taylor undertook the first mineral examination of this district in 1832. He discovered its basin form, but remained uninformed as to its western limits. But he avoided Eaton's great blunder. He says with great decision: "Northward of this mineral district it would be in vain to search for these coal strata; for independent of the geological character of that country being dissimilar, consisting almost wholly of the old red sandstone, there is no ground lofty enough to contain them in that direction, unless in the case of another insulated basin, which the structure of the whole region negatives." And yet a small patch of the lowest bed in the Blossburg basin does, as is now known, lie like a wafer on the summit of the next mountain range to the north.

But while he avoided Eaton's error of carrying the Blossburg coal slates down the Tioga river and along the shores of the New York lakes, he fell into an error of exactly opposite character, and quite as extraordinary; for he gives the dips down the Tioga river to Corning, in New York, as *all one way;* and summing up these dips, he concludes that any mountain to the northward, and near the Chemung river, "must be more than 6,000 feet high" to catch the bottom bed of the Blossburg coal measures, whereas, as he rightly observes, "they do not commonly exceed 600 feet." How so shrewd an observer could have failed to see that, descending the Tioga river, the dip is reversed four times, making two strongly marked synclinal basins, in each of which lies a mountain, and the top of each mountain capped with the conglomerate, on which lies the lowest Blossburg coal bed—is one of those inscrutable problems in the history of science over which we, in the light of the accumulated knowledge of the last half century, can only brood amazed. But the same prejudices which saw lake terraces in Kishacoquillas valley could blind the eye to dips which, although apparently reversed, ought not to be considered so in the light of his preconceived ideas.

But there is another excuse for this best of our early explorers, whose very errors must be handled with a curious respect. He calculated that the rocks in the banks of the Chemung must plunge to a depth of "6,275 feet below the summit of the surrounding hills of the Tioga (Blossburg) basin;" for, he had come fresh from the Broad Top and Allegheny mountain outcrops, whose *thickness* he expected to see represented on the Tioga river. It was several years before geologists learned how rapidly Formations XII, XI, X, IX, VIII, (Taylor's Old Red series) thin northward towards New York; so that a total thickness of 35,000 feet at Harrisburg, is represented by a thickness less than 10,000 feet between the Mohawk and the Pennsylvania State line. Yet so confident was Mr. Taylor of his conclusions, that he gives a section from Blossburg to Corning, on Plate VIII, page 194. Of course this is unfortunate, for there can be no mistaking his language when thus illustrated; and any other sections published by him in those days must be viewed with grave suspicion that his data were badly gathered, or distorted to suit his prejudices. It also renders one quite averse to listen docily to his account of "nine coal beds, varying in thickness from one to three feet, included within a space of 144 feet;" and to his estimate of 70,000,000 tons of coal as the total contents of the basin.

One of Mr. Taylor's good observations in this memoir relates to the limestone in IX, two miles north of Mansfield, the "cornstone" of English geologists; and his reference to the same limestone found in the same position by Mr. Edward Miller on the line of the Pennsylvania (Old Portage) railroad, west of Hollidaysburg, and 150 miles south-west of Blossburg. His analyses of it by I. W. Alder and by Mr. T. S. Clemson (on page 215) are interesting.

Mr. T. G. Clemson's analyses of five coals from Blossburg, are annexed to Mr. Taylor's paper.

5. "Memoir of a section passing through the Bituminous Coal Field near Richmond, Va.," followed by analyses by T. S. Clemson, (page 275.)

6. "Review of Geological Phenomena . . . in Virginia and Maryland," &c., (page 314.) In this he gives also a plate of fossil plants from the New Red near Fredericksburg, which is of interest to Pennsylvania geologists.

Mr. Taylor continued incessantly observing and describing geological facts for many years, in fact until his death in 1851. He made himself at home in the primary, as well as secondary rocks, but preferred to be employed by owners of coal lands. In 1848 he resumed all he knew of coal in a volume of 750 pages, entitled "Statistics of Coal," which was re-published in 1855, under the editorial care of Prof. S. S. Haldeman, one of Mr. Rogers' assistants on the First Geological Survey. Mr. Taylor left a mass of manuscript, materials, geological maps, beautifully drawn and colored; sections; calculations, &c., most of which are now in private libraries, scattered, but treasured by their owners, and still of considerable value.*

It may well be said, therefore, that Richard Cowling Taylor was the forerunner of the First Geological Survey of Pennsylvania, and the best of the early geologists of the United States; certainly the only geologist who entirely deserved that technical title among the members of the Geological Society of Pennsylvania.

Dr. Gerard Troost was State Geologist of Tennessee, and his three memoirs in the Transactions relate to the history of Geology in that State.

Mr. Edward Miller was a Civil Engineer of great ability, in charge of the construction of the State railway across the Allegheny mountain, constructed for taking the boats from the Juniata canal at Hollidaysburg by five inclined planes to the top of the mountain, and sending them down by five other inclined planes to the head of the Conemaugh canal at Johnstown.

The railway cuttings on the front face of the mountain permitted him to make a geological section of the Old Red sand-

* It is entirely proper to make it known that Mr. Taylor's manuscript maps of the gold regions of Virginia, &c., &c., are still in the possession of his widow, Mrs. Emily Taylor, 2043 Chestnut street, Philadelphia, who would gladly sell them to some geologist who might not only make them useful, but aid in keeping the memory of her distinguished husband alive in those parts of the United States the mineral resources of which he so long labored to discover, make public, and develop.

stone system (XII, XI, X, IX and VIII); and this he published (pages 251–256) in the Transactions of the Geological Society in 1834–'5. Had he not, in obedience to the taste of the eastern geologists, and from habit as a constructing engineer, *exaggerated the vertical scale to eight times that of the horizontal scale*, so as to distort all the dips, this section would be not only of the highest interest as a classic in the science, but would stand us in capital stead in our annual report this year (1874–'5); comparing it, as we then could, with the elaborate section which Mr. Platt has had made for his report on the Clearfield and Jefferson district, along the Snow Shoe and Bellefonte railroad. Mr. Platt will give Mr. Miller's text in his report on Cambria and Somerset counties, (Report of Progress in 1875,); but the section as Mr. Miller published it is worthless; and, alas! the suite of specimens which he sent with it to the museum of the society in 1833, cannot now, perhaps, be recovered.

Mr. Miller's lowest and first division of rocks, containing the Helderberg limestones (No. VI) and Oriskany (No. VII) is not measured; his second division, of Hamilton, Portage and Chemung rocks (No. VIII), he makes to be 5,710 feet thick; his third, of Catskill rocks (IX), measures 3,370 feet; and his fourth, or Coal Measure division, which he commences as low as No. X, no doubt to the satisfaction of our western geologists, he could not measure, partly because of the distortion of his section, and partly because of the erosion of the coal measures at the top of the mountain, and the very difficult geology descending the western side.

His most interesting observation is that of the bed of "silicious limestone, 30 feet thick" at the foot of Plane No. 7, which is the fourth ascending the mountain from Hollidaysburg, the first coal bed appearing at the head of the same plane. The foot of the plane being 1,800 and the top of it 2,061 feet above tide water; the dip being about 10°. This is the limestone which Taylor traced along the north face of the mountain at Blossburg.

Dr. Harlan published a paper on the coal plants sent down to the society by Mr. Miller, and four of them are figured in plate 14, facing the memoir. He called one species *Pecopteris*

Milleri, another *P. obsoleta*, but a fine *Neuropteris* he does not specify further than that "it is not remote from *N. macrophylla* or *N. flexuosa* of Brongiart. Mr. Miller told him that layers of rock holding marine shells lay over and under the shales with plants; then he goes on to say: "The anthracite Coal Measures of the Lehigh and Schuylkill, are generally referred by geological observers to the grauwacke series; and the bituminous Coal Measures of Allegheny, Ohio, &c., to the secondary formations—the rocks would lead us to the former opinion, the fossils in some instances to the latter, so far as they have been examined from both localities."

But fossil botany must not claim to have dispersed this error; for it clung to the minds of geologists long after the topographical and structural studies of Frazer, Jackson, M'Kinley and Henderson, without any aid from fossils, made it a ridiculous superstition of the past.

Mr. T. A. Conrad also communicated a paper respecting Mr. Miller's Portage railroad shells, and in the accompanying plate 12, he gives drawings of five species which he names with some hesitation: *Turbo tabulatus*, *T. insectus*, *Stylifer primogenia*, *Pecten armigerus* and *Productus confragosus*. But he avoids committing himself to the grauwacke theory, although he alludes to it. Two years afterwards he was disproving it himself in the capacity of palæontologist of the New York Survey.

Mr. Clemson's analyses of Mr. Miller's specimens of Allegheny mountain coal follow next in order in the Transactions of the Geological Society of Pennsylvania, and show how crude were the ideas of chemists then, respecting the demands which a future coal trade would make on their art; not a word about sulphur, phosphorus, potash, soda, magnesia,—merely volatile matter, 15; ashes. 8; carbon, 77=100.

Dr. Harlan read another short paper on a fossil plant (plate 14, fig. 4) from an anthracite coal mine, which he names *Equisetum stellefolium*, but which he evidently suspects may be an *Asterophyllites* or *Annularia*.

Mr. H. Koehler read a paper "On the anthracite deposit at Tamaqua, Schuylkill county, Pa., with a map and section," (p. 326–328,) and it is a most creditable cross section of the Coal Basin along the Little Schuylkill, showing great local knowledge, and how thoroughly the different beds had been opened previous to 1834 '35. Even the sharply folded synclinal at the base of the Sharp mountain is signified; but the anticlinal lying more towards the centre of the basin is not noticed. The vertical and horizontal scales are the same, the contouring excellent, and the accompanying map is well lined.

Of the remaining eight or ten memoirs published in the transactions of the Geological Society of Pennsylvania, none relate specially to Pennsylvania geology.

The society is known to have contemplated a survey of the whole of Schuylkill county by a committee of its members; for in 1833 part of the necessary fund had been subscribed, and landholders in the coal region were invited to complete the fund required. But the work was never accomplished until 1837, when the State Geologist gave it in charge to Dr. Whelpley and Mr. Sheafer.

Another piece of work by the secretary of the society, Mr. P. A. Browne, namely, a careful set of observations along the Schuylkill river from its mouth to the northern line of Montgomery county, was never published.

Two of the most active members of the society, Mr. T. C. Conrad and Mr. Isaac Lea, occupied themselves with the Tertiary fossils of New Jersey, Delaware and the Southern Atlantic and Gulf States, and these gentlemen have published a large scientific literature on this subject and continue still to do so. But the Tertiary Formations are not represented in Pennsylvania, except, perhaps, by the Lignite deposit of the Pond Bank in Franklin county, which, in this, resembles that of Brandon in Vermont, and deserves the most minute investigation.

While this memoir was going through the press Professor Prime reported the discovery of a nut and other evidences of a similar and possibly Tertiary deposit in the excavations of the old Balliet mine at Ironton west of Catasauqua, in Lehigh county.

This discovery will re-open the question of the recent age of our brown hematite ore beds, so confidently affirmed by the late Dr. Hitchcock, State Geologist of Massachusetts and of Vermont. It will stimulate the observation and publication of similar occurrences in other ore banks between Vermont and Alabama. It will also have its due weight in the discussion of the part which glacial ice has played in collecting, scattering and depositing fragments of such Tertiary lake deposits as *may* (let us not forget the possibility) have once lain upon the old and now eroded surfaces of our Appalachian valleys; although no trace of such has as yet been noticed, except one at the salt works west of Wythe, in Virginia. But such discoveries must not blind us to the accumulated evidences of the *genetic* connection of these brown hematite ores with the Devonian and Silurian (or Cambrian) rocks along the outcrops of which alone they occur.

CHAPTER III.

A HISTORY OF THE FIRST GEOLOGICAL SURVEY OF PENNSYLVANIA.

The act of Legislature appointing a survey of the State was dated the 29th of March, 1836, and authorized an annual appropriation of expenditure of $6,400 for five years, to pay the salaries of a geologist, two assistants, and a chemist.

Professor Henry D. Rogers was appointed geologist, Mr. James C. Booth and Mr. John F. Frazer, assistants, and Dr. Robert E. Rogers, chemist.

The first season's field work sufficed to make known with certainty the geological order of the rocks of Middle Pennsylvania; and on this determination, as a sure foundation, all subsequent work in the Appalachian mountain belt of the Atlantic States was based. Until then the mountains of IV (Oneida) and of X (Catskill) had been confounded together. But when Mr. Frazer made his descending section along Yellow creek in Huntingdon county, from the Broad Top Coal Measures at Hopewell, through the gaps in Terrace mountain (X) and Tussey mountain (IV), to the Lower Silurian limestones in Morrison's cove,—and Dr. Booth reported the same unmistakable order of formations on his return from a tour round by the Potomac,—we may consider that then the general geology of Pennsylvania and of the Atlantic States was settled. It has, in fact, suffered no modification of any great importance in stratigraphy since that time.

Yellow creek (with its prolongation Sandy run) will always have a classical interest for American geologists. It is one of the few lines of continuous one-dip section, straight across the Old-Age System, through all its slanting formations, from the coal measures at its top to the magnesian limestones near its bottom; with a total thickness of rock deposits of about twenty-five thousand feet.

Professor Frazer was fond of telling his geological friends the story of this adventure, explaining how, up to the rendezvous of the geologist and his two assistants at Huntingdon one summer Sunday, their explorations had been a series of embarrassments; their note books filled with a confused mass of irreconcilable statements; their cross-sections contradictory in themselves when compared together; and the geology of Pennsylvania apparently at variance with that of the neighboring State of New York. His Yellow Creek section was, however, disbelieved, until verified by Prof. Rogers himself the following week, and by Prof. Booth afterwards on the Potomac. From that moment everything went smoothly; all contradictions vanished; their back notes became luminous, and the northern outcrops of Hall and Vanuxem, in New York, were seen to be all represented in the same regular order, although on an immensely enlarged scale, by the southern outcrops of the same formations in Pennsylvania; Tussey mountain being made by the Oneida conglomerate, No. IV; and Terrace mountain by the Catskill sandstones of New York, Nos. IX and X.

Prof. Rogers was able to recognize, in co-operation with his equally distinguished brother, the State Geologist of Virginia, the persistency of the same formations, under slight variations of color, size and mineral ingredients, across the Old Dominion and into Tennessee and Alabama; and therefore, we may truthfully claim for the work of the first year of the Pennsylvania Geological Survey, 1836, that it gives the epoch for American Old-Age Sedimentary Geology.

The Palæozoic system was divided into twelve parts or formations, the lowest being that the limestone of Harrisburg and Reading, the highest being that of the Coal Measures. This earliest classification, modified somewhat in the following year affixed the following numerals to the formations:

XII. Coal Measures.
XI. Red shale.
X. White sandstone, } of the Second Mountain.
IX. Red sandstone, }
VIII. Olive shales, &c.
VII. Cherry sandstone.
VI. Limestone.

V. Red shale and fossil ore.
IV. Gray sandstone,
III. White sandstene, } of the First Mountain.
II. Slate,
I. Limestone, } of the Lebanon Valley.

No modification of this system of numbers would have been needed had the survey been commenced on the Schuylkill, instead of along the Juniata. But no one can take exceptions to the plan of the first reconoissance because it assumed the fact that a long section through the central district of the State would be the one most likely to reveal the structure in all its completeness. It could not be forseen that all the elements exhibited along one hundred miles of one river, the Juniata, were compacted along twenty miles of the other, the Schuylkill; nor could it be anticipated that formations thousands of feet thick on the Schuylkill would be found dwindled away to hundreds on the Juniata.

It was glory enough for the first year's survey to have swept away the enormous fictions of previous years,—demolished the theories of Taylor respecting the Broad Top and Anthracite Coal Measures,—put a stop to all talk about "Transition Rocks,"—and separated the mountains of IV and the mountains of X into two distinct groups of very different ages.

We may with equal truth assert that American Structural Geology was born in 1836. For it was during this first season's work that that system of anticlinal and synclinal folds, or rock-waves, was first fairly seen and understood by Professor Rogers, with which his name will ever be identified as closely and as honorably as the names of Thurmann of Switzerland, and Elie de Beaumont of France,—although, it was not until the results of subsequent seasons' work were plotted to scale, that he had the opportunity of demonstrating the normal shape of these waves, and the evident movements of the earth-crust towards the north-west, which their almost uniformly steeper dips towards the north-west exhibited.

The second year of the first Geological Survey was 1837.

The act of Legislature of March 29, 1836, was amended in

the spring of 1837, by enlarging the appropriation to allow of four assistants instead of two. These were Mr. Samuel S. Haldeman, Mr. Alexander M'Kinley, Mr. Charles B. Trego and Mr. James D. Whelpley.

The appropriation for incidental expenses permitted the employment of four aids or sub-assistants: Mr. Alfred F. Darley, Mr. Edwin Haldeman, Mr. Horace Moses, and Mr. Peter W. Sheafer.

The two assistants of the first year had resigned, to assume important positions, which they held with honor to themselves and advantage to the community. Mr. Booth became Chief Geologist of the State of Delaware, and afterwards Chemist to the United States Mint. Mr. Frazer became Professor of Natural Philosophy and Chemistry in the University of Pennsylvania, and editing secretary of the Franklin Institute, Secretary of the American Philosophical Society and member of the National Academy at its incorporation in 1861. His death in 1872, was universally mourned. Dr. Robert E. Rogers remained chemist of the Survey.

The expenses of the first year had been but $2,700. Those of the second year amounted to $6,500; a sum equivalent to fifteen or twenty thousand dollars now. More would have been spent, but only two assistants took the field in the spring; the other two were not selected and commissioned until July. In fact, nothing could be more difficult, forty years ago, than the discovery of persons of a scientific education, suitable for such employment. Of the four assistant geologists of 1837 Mr. Whelpley alone had had a geological training, under the elder Silliman, of Yale College, Connecticut; Mr. Trego had paid some attention to botany, and Mr. Haldeman to the fresh water shells of the Susquehanna river, which flowed before his homestead. Mr. Sheafer had grown to manhood in sight of the old Lykens Valley collieries.

In the first year of the Survey a good general idea of the order of the principal formations had been formed, and a tolerable sketch made of the direction of their outcrop belts across the State. Mr. Rogers now felt himself in a condition to take up the survey in detail, and resolved to devote the entire season of 1837 to one quarter of the State, bounded by the Delaware,

Lehigh and Susquehanna rivers and the New York State line; Easton, Harrisburg, Tioga and Great Bend being at or near its four corners. This district contained none but the sedimentary, fossiliferous formations of the Old-Age system; and it contained them all, from the top of the Coal Measures, at Pottsville, to the bottom of the Silurian, at Allentown, Bethlehem and Reading. It confined within its limits all the Anthracite coal basins, and on these was fixed the attention and were expended the efforts of the Survey in 1837.

The Pottsville basin was very carefully surveyed by Mr. Whelpley, and all the then open mines examined and described. Straight lines were run with compass and chain across the Beaver Meadow mountain basins, and across the Wyoming Valley basin, by Mr. M'Kinley. Mr. Trego studied the Lehigh country, and Mr. Haldeman the country of the Middle Susquehanna.

Three discoveries of importance, resulting from this second season's work, changed the numbering of the formations (from I to XII) in the following way:

1. A sand-rock was discovered at Bethlehem, Allentown, Reading and Chiquesalunga, no where visible at the surface in Central Pennsylvania; underlying the great limestone valleys; and corresponding to the Potsdam sandstone of Lake Champlain. This Mr. Rogers now had to call No. I. The limestones (I) became therefore No. II; and the slates (II) became for the same reason No. III.

2. The double crested and terraced mountains of middle Pennsylvania: Bald Eagle, Nittany, Tussey, Standing Stone, Jack's, Black Log, &c., made by the outcropping of two massive sand-rock formations, (III and IV,) with an interposed softer sand-rock formation, were found to become, when followed across the Susquehanna into Eastern Pennsylvania, a single crested mountain, (called North, Blue, Kittatinny,) composed mainly of one sand-rock, or a series of alternate sand and mud-rocks. The two sands, III and IV, of the last year of the survey, with their enclosed rocks, were, therefore, united into one formation, and called simply No. IV. From that time to the present we have spoken of the "Upper, Middle, and Lower of IV."

3. The above changes left the numbering of the formations above IV unchanged. But by this year's work another discovery was made: that the lowest sand-rock of the Coal Measures (XII,) which, in Middle Pennsylvania, had been thin enough to be disregarded, but was on the contrary one of the great formations of Eastern Pennsylvania, deserved, and in fact demanded a separate number of its own in the series. From 1837, then, down to the present time, we speak only of the Great Conglomerate, the Millstone Grit of England, as No. XII. Mr. Rogers added a No. XIII to signify all the Coal Measures lying above the Conglomerate. Subsequently this was found to be a useless and embarrassing addition, and was abandoned by him. In truth, if XIII were to be retained for the (Lower) Coal Measures immediately above the Conglomerate, we should call the Barren Measures XIV, and the Pittsburg (Upper) Coal Measures XV, and the Greene and Washington County (Upper) Barren Measures XVI. For these formations differ from one another as truly as do VIII, IX, X and XI, from one another.

Mr. Rogers showed good judgment in employing at first numbers instead of names to distinguish the great formations; at least until the sameness of our formations with those of New York and those of Europe could be proved; until he could be sure that no formations named elsewhere were absent here; until he was certain that Pennsylvania possessed no peculiar deposits of her own which existed nowhere else; in a word until all the rocks in all parts of the State had been found, examined, measured and placed in order. It would then be time enough to name them all according to a system. Meantime their order in time would be best stated by their several numbers; No. I being the oldest and lowest, and No. XII the youngest and highest in the series. And these numbers might be diminished, increased or shifted to suit the discoveries of each succeeding year of the survey.

But, unfortunately, numbers are of the essence of names in one respect: they stick as tenaciously to the things they represent. Mr. Rogers could only slightly change his numbering even after so short a time as the one year elapsing since its adoption. With these slight changes the numbering of 1837 set so firmly that it never afterwards was changed; and it will probably last as long as our science. Its convenience, indeed, has been demonstrated by thirty years of use; and it stands a monument of the good work of the first two years of the First Pennsylvania Survey. All the principal features of the Old-Age system of rocks was, in this system of numbers, painted with a bold, free brush, truthfully; and no good reason can be given for its non-adoption by other State surveys, except that some of the principal formations, so grandly exhibited in Pennsylvania and Virginia, are miserably thinned down and impoverished, or disappear entirely, when followed into other districts of the Continent.

Had Mr. Rogers not adopted another nomenclature when he published his final report,* or had a final report of the Geology of Virginia been published, retaining this original numbering of its annual reports, then, doubtless, Mr. Safford would have adopted the same numbering for the same formations in Eastern and Middle Tennessee, and western geologists would have been compelled by so large a published literature to follow the same method.

But after all said and done, the inherent objections to numbering the rocks remain potent, and oblige all geologists to employ in their descriptive geology another system: that of geographical names. Among the twelve great formations of Pennsylvania there is no place found for one of the most important of the groups of New York and Upper Canada: the Niagara. This formation, which makes at its northern outcrop so mountainous a mark on the shores of Lake Huron, is too thin to be discovered at its southern outcrop in our State. If it exists at all, it forms some of the small layers of our No. VI. Again; our No. 1, which is but 30 feet thick upon the Lehigh, is said by Professor Cook to be 3,000 feet thick in Green Pond mountain in New Jersey; and Mr. Safford separates it into two huge formations, the Ocoee, 10,000 feet thick, and the Chilhowee, 2,000 feet thick, in Eastern Tennessee. If Dr. Newberry had adopted numbers instead of names in his Geology of Ohio, he would have found himself in the same embarrassment, and even greater.

*See his annual address before 4th Albany meeting of American Naturalists and Geologists, April, 1843, page 20 of his address. Idea formed in 1840, "three years before."

For our No. I and the greater part of No. II do not appear above the surface in that State; and No. X is not recognized by him as existing there, although in Eastern Pennsylvania it is 2,000 feet thick. On the other hand, the Niagara is well developed in Ohio.

Geographical names for rock strata become therefore a necessity. This necessity increases moreover as geology advances, and as better study of the great formations enables geologists to divide and sub-divide them into groups and layers. Sometimes a single layer, only a few feet or a few inches thick, can be traced for hundreds of miles by its peculiar fossils, or by its peculiar mineral value. Many such may occur in one formation. Each must have a name. If letters be used for this purpose, (the plan pursued by Mr. Safford,) I*a*., I*b*., I*c*., II*a*, II*b*, II*c*, &c. then the old difficulty is encountered in a slightly different shape, for it will be as hard to shift the letters, when new layers are discovered to require them, as to shift the numbers in a well established system. Then some cumbrous "third term" system must be adopted, such as: I*a*, Ib^1, Ib^2, I*c*, &c., or, Ia', Ia'', Ia''', &c.

We are driven to the use of geographical names. Although here a difficulty of a wholly different order meets us. The rock may not be the same in one place as in another, and the name of one rock may be given wrongly to another. Chiques Rock, on the Susquehanna river above Columbia, may *not* be Potsdam sandstone in northern New York, and at Bethlehem and Allentown. But then, neither will it be No. I. So, the name of one fossil shell, or plant, or fish, or mineral, may be falsely bestowed upon another. But when the mistake is discovered by comparing many specimens, by finding them more perfectly preserved, by dissection, by microscopic inspection, or by analysis, the mistake is easily rectified, and a name of its own can then be given to the later found fossil. Just so, if Chiques rock be not Potsdam, why then it remains simply Chiques, and not Potsdam. That is precisely the advantage possessed by geographical names; they can wait patiently any length of time for rectification. They show also where such and such a deposit has been first best studied, and where it makes so strong a mark on the earth's surface as to attract notable attention.

It is an accepted law in science that an organic form once named by its discoverer, if its name can bear the test of long continued criticism, pre-empts that name, and maintains its claim to ownership of it for all time, against all innovators. The law holds good in the case of mineral beds and rock layers. They are sedimentary *formations*, i. e. organized beings, which have been born at a certain point of time, grown in an ascertainable shape, to a measurable size, through a limited age, and were then overgrown, buried, fossilized for our contemplation. Each has an anatomy which can be dissected, a constitution which can be analyzed, distinguishing it from others. The geologist who first studies and names it correctly has a right that his name for it should be accepted by all who approve of his statements respecting it. Disputes may arise. Two geologists may, at the same moment, give different names to the same formation in different districts. But this rarely happens; and popular usage will be sure to select one of the two names in preference to the other, or fuse them into one, as the Irish geologists have done, in calling a certain system the Cambro-silurian system, which Sedgewick claimed to be the upper part of his Cambrian, and Murchison insisted was the lower part of his Silurian.

The most recent literature of our science proves that there is no getting rid of the old geographical systematic names, such as Jurassic, Devonian, Silurian, Cambrian, Huronian, Laurentian. They are historic fixtures, like the higher titles of the nobility and rich-landed gentry of Europe. Their circle is exclusive. It is exceedingly difficult for a newcomer to get in, even after the purchase of a part of their estates. It is doubtful if Dr. Hunt's Norian System will "make a family," or Professor Dana's Canadian Period, or Dr. Newberry's Erie Epoch. For the new names are invaders of the old. A name may be excessively local at first, but age and use may give it continental occupation. As the imperial house of Hapsburg was cradled in a little castle on a secluded cliff, but came to rule half Europe, so Potsdam is the name of a village hid behind the Adirondac mountains on the borders of Canada, but has become the accepted title for the foundation of the Old-Age System in all parts of North America, where Geologists have found its characteristic worm burrows and trilobites. And it is a name which resists all attack because there is nothing in it to attack. In Professor Rogers' revised nomenclature, it is called the Primal sandstone. But that begs the question of its being actually the bottom formation of the unchanged, fossil-bearing sediments. In Virginia and Tennessee there seem to be many sand-rocks below it. Then it is not Primal. But it remains Potsdam.

The name Carboniferous System is another example of the inconvenience of any but geographical names. It is now well made out that there are *several* carboniferous systems, of very different ages, in many places of the earth's surface. Even in Great Britain, where the name was first invented, it is now clearly proven that the Carboniferous System of Middle England and South Wales is not the same with the Carboniferous System in Scotland. The Coal Measures of Middle Virginia and East Tennessee do not correspond with those of Western Pennsylvania and North-Eastern Ohio. A world of Carboniferous rocks has been discovered in India and Australia and China, which appears to be not at all of "Carboniferous" age. One of the great discoveries of the First Geological Survey of Pennsylvania was, that the Upper Coal Measures of the Monongahela river and the Lower Coal Measures of the Allegheny river, were two quite different and widely separated systems. Then followed the discovery, by the Virginia State Survey, that both these systems were different from and far separated (in time) from that of the New river (or Upper Kenawha.) Then followed the discovery of a fourth and still older Coal Measure System at a great depth (geologically) below all three, that of the Juniata river in Middle Pennsylvania.

Without being fully aware of all this, but reasoning from what they saw in the Western States, the western geologists came to a pretty unanimous determination to extend the carboniferous system down so as to take in almost the whole Devonian system.

Dr. Dawson has remarked on the amount of graphite in the Oldest Age System (Laurentian) as so enormous, that Sir William Logan was justified in thinking that the quantity of vegetable life then flourishing upon the planet may have equalled that of the marshes and forests of the "Carboniferous age." If this be so,—and Lesquereux speaks in the same tone,—we might with equal propriety absorb into the "carboniferous" system, not merely the Devonian, but the Silurian, Huronian and even Laurentian systems. With the same propriety we might include in it the Trias, because of its North Carolina Coal Measures; and the Cretaceous and Tertiary systems, because of their Rocky mountain coal beds. All the formations of the world from the lowest to the most recent, would then become "carboniferous." However ludicrous this *reductio ad absurdum* may appear, it is a logical consequence of the method pursued by our western geologists, when they carry down the Coal Measures to the base of the Devonian system, and obliterate the word Devonian from the column of formations,—for very different reasons, be it remarked, from those which influence the new school of English geologists in their endeavor to expunge it from their lists.

To the geologists of Pennsylvania and Virginia, all this was a palpable absurdity. But the temptation to commit it would not have occurred, had we at first adopted geographical names for our various Coal Measures; had we established a Monongahela coal system, an Allegheny coal system, a New River or Upper Kenawha coal system, and a Juniata coal system (in the Devonian.) A geographical name gives rise to no contests other than of identity; no conflicting classification; and therefore it need never be changed. It says: "There lie the rocks; call them what you please."

Names, partly geographical and partly descriptive, were first assigned to the principal Old Age Deposits of the United States by the geologists of the State of New York, Mr. James Hall, Mr. Lardner Vanuxem, Dr. Ebenezer Emmons, and Mr. W. W. Mather; and that as early as the year 1836, which was the first year of the New York survey. The first party took the field about the middle of June; others as late as September in that year. The names provisionally proposed for the beds were discussed by the geologists together in the winter. A few were adopted in the winter of 1836–'7; more in the winter of 1837–'8; at the end of the season's work of 1838, the New York geologists were pretty well agreed upon the names of the formations to be rejected and adopted. They then began to consider the necessity of a conference with the geologists of other States. But a conference with Prof. H. D. Rogers, of Pennsylvania, and W. B. Rogers, of Virginia, failed to produce a common system of names for these three important States. The New York names were used in the New York progress reports of 1837, 1838, 1839 and 1840.

These names have become classical; are taught in schools; are employed in the reports of geological surveys of other States; are attached to specimens of rock, and to fossils, in all museums. There seems to be no good reason for abandoning their use, or for refusing to apply them to the same beds of rock wherever they come to the surface, and can be recognized.

It is quite true that they do not form a consistent, ideal system of names; some of them being geographical, while others are names taken from fossils, or minerals which give character to special beds.

Of the first kind are:—Catskill, Chemung, Portage, Genesee, Tully, Moscow, Ludlow, Marcellus, Schoharie, Oriskany, Onondaga, Niagara, Clinton, Medina, Schwangunk, Hudson River, Utica, Trenton, Black River, Chazy, Potsdam.

Of the second kind are:—The Enerinal limestone of the Hamilton group; the Corniferous limestone; the Candagalli grit; the Upper and Lower Pentamerus limestones, and Delthyris Shaly limestone between them; and the Birdseye limestone of the Trenton group.

Of the third kind are: the Coal Measures, the Sub-carboniferous limestone, the Water lime group, the Salina (Onondaga salt) group, the Cement beds, the Gypsum beds, the Calciferous sandrock.

This mixture of place-names, fossil-names and mineral-names in the one grand system of the Old-Age formations, by the New York geologists, was a great convenience to them, and has been a great convenience to the geologists of other States. Nor is it any more flagrant departure from consistency than the English geologists have committed in naming the great systems together, thus: (beginning at the top) the Tertiary system; the Cretaceous system, named from its mineral chalk; the Neocomian or Lower Cretaceous system, a name adopted from the

French; the Oolite system, named from the fish-roe structure of its principal mass; the Lias, so named by the common people of a part of England—the Oolite and Lias being called the Jurassic system, a geographical name; the New Red, named from its color, or the Triassic system, so named because divisible into three; the Permian system, a geographical name; the Carboniferous system, named from its coal beds; the Old Red system, named from its color, and considered to be the same as, or the upper part of, the Devonian system, named from Devonshire; the Silurian system, named from the race of savages which inhabited the west of England in the time of the Romans; the Cambrian system, named after the ancient Welsh; the Primary system, so named because it was once accounted by geologists to mark the beginning of geological history, and to be the foundation floor of the surface rocks of the earth.

This is a heterogeneous list; but it has served geology very well, and continues in use wherever geologists work, teach and talk together. None of the fanciful arrangements of names all of one kind—and many have been invented and published by distinguished geologists—has ever succeeded in banishing or suppressing these good old English names, over the exact meanings of which many a battle has been fought.

The fact is: things, like people, are named by accident, christened under excitement, or designated for some peculiarity; and it is against a natural law of the human mind and contrary to the habits of mankind, to change the name when once it has become notorious. Legislation is of no force against old names. The City Councils of Philadelphia kept up signboards marked "Mulberry" and "Sassafras" and "High," at all the corners of the three streets so named by William Penn, for twenty years; but the citizens had from the first chosen to nickname these streets Arch, Race and Market, and continued so to call them all those twenty years, and call them so still. The useless signboards were taken down and the peoples' names mounted in their places. Nor will all the efforts of certain geologists to introduce such Continental names as Neocomien, Dyas, Rothe-todtliegende, &c., avail the weight of a feather against Anglo-Saxon obstinacy, calling things by the English names which first strike the imagination as characteristic, no matter what learned or what æsthetic system of nomenclature may happen to be violated.

If, therefore, the New York rocks were well named and early named where they sink to descend, southward, under the southern counties of that State and the northern counties of our own, then, it is not only reasonable but it is inevitable that they will be recognized by the same names wherever they ascend to the surface in Middle Pennsylvania. No artificially constructed set of new names can come into common use. For the great object of geology is to harmonize facts, to identify the rocks of distant localities; and their study proceeds from district to district, carrying the names of the districts studied first to fix them in the districts to be studied next. The geologist of Pennsylvania has one question forever on his lips, when he meets a geologist of New York: Is this rock of mine your Potsdam, your Niagara, your Waterlime; or is it different? When he meets a Canadian geologist the first question he asks him is this: What rock in Pennsylvania must I call "Laurentian?" Is our South Mountain "Huronian?" Am I to name our Philadelphia belt "Quebec Group;" or what else? The instructed discoverer always gives names to the uninstructed searcher. The New York geologists were ahead of those of Pennsylvania in *naming* the Old Age rocks; therefore, all the Old Age rocks in Pennsylvania recognized in New York must wear New York names. Other rocks in Pennsylvania not reaching into New York, or for any reason not named there, must have Pennsylvania names given to them. In like manner such Ohio and Michigan names as "Waverly" and "Marshall" will probably be accepted and welcomed into the Pennsylvania list, because they stand for groups of beds not distinguished with sufficient care by the First Survey of our State, but admirably studied and described by Prof. Winchell, Dr. Newberry and their coadjutors. And it is quite possible that some of Prof. Safford's Tennessee names will find themselves at home in Pennsylvania; and some of Prof. Kerr's North Carolina names will fasten themselves firmly to the rocks of our south-easternmost counties.

It is none the less true that the numbers adopted by Prof.

Rogers, and used in all the annual reports of the First Survey, were so just, and agree so well with the great features of the surface, that they can be still conveniently and usefully employed as sub-divisions of the English systems: I, II, III representing the Cambro-silurian; IV, V, VI, Silurian; VII, VIII, Devonian; IX, Old Red; X, XI, Sub-Carboniferous; XII, Carboniferous. If some geologists prefer to call IV Middle Silurian, VIII, IX Devonian, and X, XI, XII Carboniferous, it will make no difference at all in respect to our knowledge.

The subject of the nomenclature adopted by Mr. Rogers for the Final Report, and the subject of a geographical nomenclature possibly suitable to all the States of the Atlantic sea-board will come up for discussion in their proper places. But that this year of the survey under consideration (1837) saw for the first time a consistent, general and permanent classification of the Old Age rock system of North America, cannot be better stated than in the words with which Mr. Rogers begins his chapter of General Observations on the results of the season's work, in his report to the Secretary of the Commonwealth, dated Philadelphia, January 27, 1838, pp. 82, 83:

"In taking a general review of the extensive series of our Appalachian formations, now for the first time systematically classified and described, our attention is forcibly arrested by their vast thickness, the immensity of their range, and the inexhaustible stores of mineral treasure which they contain.

"From the base of the entire series, where the bottom of the lowest sandstone is in contact with the primary rocks of the South mountain, to the uppermost beds of the Anthracite coal measures, the absolute depth of this enormous group of strata in our counties east of the Susquehanna, cannot be less than *forty thousand feet.* It is worthy of remark, that probably no other district in the entire Appalachian chain, from the Hudson river to Northern Alabama, presents our American lower secondary rocks on an equally expanded scale, or so admirably developed for geological investigation.

"The gigantic magnitude of the areas covered by these thirteen formations, may be conceived, when I state that they not only occupy the entire surface of Pennsylvania, with the exception of the corner of the State, south-east of the South mountain, but that with a few interpolated strata, they comprise three-fifths of the territory of the United States, east of the Mississippi.

"In an essay still unpublished, but written nearly a year since, based in part upon my own personal observations, and in part upon a comparison between these and the numerous insulated descriptions of our rocks given by various Geologists and travellers, I have attempted, and as I believe successfully, to trace individually the formations of our great Pennsylvania series south-westward along the mountains as far as Alabama, and also to identify them in their course across New York and the north-western States and Canada, to the northern shores of Lake Huron and Lake Superior.

"Within the whole of this wide expanse of country, researches will develope I conceive but *a single*, though vast group of strata, the successive sediments of one immense ocean, the creations of but one prolonged Geological epoch, commencing almost in the dawn of marine animal and vegetable existence, and terminating with the latest produced deposits of the coal.

"Viewing the majestic scale of our formations, and the combined grandeur and simplicity of structure of the enormous geological basin which they embrace, we turn with grateful satisfaction, to the peculiar position which Pennsylvania occupies in this vast area. Lying on the margin of the great Secondary basin of the United States, and traversed as it is, for nearly three hundred miles through its centre, by the whole broad belt of the Appalachian or Allegheny chain, in which a system of gigantic anticlinal elevations brings the entire series of formations several times in succession to the surface, it holds, in combination with Western Maryland, the *key* to the geology of many of the other States, where but *a part* of the same strata are spread out in a nearly horizontal attitude, and exhibited in but a single belt."

The third year of the First Geological Survey of Pennsylvania was 1838.

The annual report of the Chief Geologist was handed in to the Speaker of the House of Representatives by the Secretary of the Commonwealth, under date of February 19, 1839. It made 118 printed pages without illustrations. But in nowise represented the amount of admirable work accomplished in various parts of the State that year. The corps was increased, and the expenses rose to $12,000, a sum equivalent, practically considered, to the present annual appropriation of the Second Survey.

Nine assistants had as many districts allotted to them, severally; and Mr. Martin Boije was appointed assistant chemist with Dr. R. E. Rogers. Mr. S. S. Haldeman had resigned, and his vacant post was occupied by Mr. Harvey B. Holl, of England. Mr. James T. Hodge, of Plymouth, Mass., Dr. R. M. S. Jackson, of Blairsville, Pa., Mr. John C. M'Kinney, of Pittsburg, and Mr. Townsend Ward, of Philadelphia, were newly appointed. Messrs. Darley, F. Haldeman and Moses were not employed.

Mr. Boije examined the country between the Delaware and Schuylkill rivers. *First district.*

Mr. Holl traveled between the Schuylkill and Susquehanna rivers. *First district.*

Mr. Whelpley and Mr. Sheafer continued their instrumental work for mapping the Southern and Middle anthracite fields. *Second district.*

Mr. M'Kinley and Dr. Jackson studied the labyrinth of mountains between the Susquehanna and Juniata rivers and the Allegheny mountain. *Third district.*

Mr. Trego and Mr. Ward wandered through the wilderness south of the Clearfield turnpike, as far as the Maryland line, (keeping east of Chestnut Ridge) to get some general idea of its geology, for its future exploration in detail. *Fourth district.*

Mr. M'Kinney explored the Monongahela and Kiskiminetas country, between the Chestnut Ridge and the Virginia State lines. *Fifth district.*

Mr. Hodge had assigned to him the Allegheny river country, from the west line of Armstrong and Venango counties to the Ohio State line and Lake Erie. *Fifth district.*

Mr. Rogers spent all the time he could spare for personal field work in the northern counties, and as far south as Clearfield and Centre counties; a country at that time an almost unbroken wilderness, roamed over by the elk and the panther.

His object was to organize for the ensuing year a camp survey of this *Sixth district.*

Most of this year's work was mere reconnoisance. In spite of what had been done in 1836 and 1837, the local geology of the State still lay quite in the dark.

The discoveries of the first season (1836) could all be summed up in the statement that the twelve great formations outcropped in a certain fixed order along the Central belt of the State; that the Anthracite regions lay in the eastern counties; tnat the bituminous coal measures covered the half of the State lying back of the Allegheny mountain; that the New Red rocks made a belt through Berks, Bucks, Lancaster, York and Adams; that wholly different, older and unrecognizable rocks occupied the South mountains; and that still other and equally undeterminable formations occupied the south-eastern corner of the State.

In the second year (1837) the work had been spent chiefly in getting on foot an elaborate survey of the most important region in the State, that of the Anthracite coal fields.

In the third year (1838) the Chief Geologist found himself face to face with the greatest difficulties. The survey of the Anthracite country had shown itself to be no child-play. The region proved to be excessively complicated in structure, and difficult of representation. Mr. Whelpley's rare genius had, indeed, a worthy field for the display of its powers. But it was evident that several successive years would be required for the study and description of the region. Meanwhile, the large iron interest of the middle counties, and the bituminous-coal and charcoal-iron interests of the western counties deserved and demanded the earliest possible attention. The northern counties also sent their representatives to Harrisburg; and these might well complain that their own part of the country had been neglected, although it especially needed all that a State survey could do in its behalf.

In 1838, therefore, Mr Rogers districted the entire State systematically, deploying his small corps over a field longer and wider than could be occupied by ten times their number. He retained his two best trained men at their post in the anthracite; and the rest traveled on horseback, by wagon and on foot, up and down all the main roads, across and along the mountain crests, up and down the river banks, collecting all kinds of geological information, visiting all kinds of quarries, mines, furnaces and natural curiosities, guided by the people of the State; with their eyes open to note all that bore directly or indirectly upon those problems of structure which even then began to make themselves apparent to their minds.

The year 1838 was the *training year* of the First Survey. At the end of it the first really good idea of the detailed *geographical geology* of Pennsylvania was grasped by Mr. Rogers and his assistants. By the end of it three or four excellent geological field hands had trained themselves for future work. Each had to struggle with all the problems of the science alone and unassisted; without books, without counselors, surrounded by thousands of people who knew where things could be looked at, but not where they could be looked for, ignorant of what the things meant, and longing to be told. It was a great school; and it made some good scholars, as the subsequent years of the survey proved.

The system of localizing the geologists in districts, the new feature of this year, and his reasons for adopting it are given by Mr. Rogers on pp. 9, 10 of his Third Annual Report in the following words:—

"Many reasons suggested the propriety of distributing the corps of assistants in the manner above indicated. It was desirable to afford to every part of the Commonwealth, as far as practicable, the advantages of an early survey of its own peculiar mineral resources, and a simultaneous exploration of several districts is obviously the only mode by which, in a limited number of years, the requisite degree of precision can be attained. Each district has a more or less different class of deposits, disposed in a special order, not to be witnessed elsewhere; it has also a different topography, which when adequately studied, affords oftentimes a valuable clue in tracing its mineral deposits; while on the one hand, it has its peculiar difficulties that demand much local knowledge to surmount, and on the other features calculated to lead to useful discoveries if familiarly understood. By allotting, therefore, a particular range of country, as far as possible, to each individual in the corps, and restricting his explorations to that alone, until he shall have mastered its geology and mineral resources in all their local details, an important economy of time is effected, and increased accuracy in the results is insured."

Had it been possible to carry out the system by retaining all the assistants from year to year, and retaining them from year to year in their respective districts, most of the embarrassments of the First Survey would probably have been avoided. But it was not possible; and is likely to be always impossible. Sickness and death make changes in all human plans. Men cannot help resigning less lucrative and certain employments to accept higher salaries, or to engage in permanent business. Men will weary of the wilderness, and of wandering in it. The loss of one employé will alter the arrangement of an entire corps. Few of the geologists of 1838 occupied the same districts in 1839 and 1840.

What the scientific and practical gains of the third year of the First Survey were can be more easily imagined than described. Every place was new; every detail of information a discovery. Not that the citizen farmers did not know the facts; for Indians and white settlers, fishermen, hunters and lumbermen, wandering Welsh miners searching after copper, tin and iron veins, had made the discoveries, and the farmers themselves had had knowledge of the discovered facts long enough before. But no one had systematically studied the facts; none had put this and that together. What each one knew he knew for himself alone, and only gossipped about at the village bar-room; it was a mere matter of curiosity or wonder, or ignorant speculation, or absurd upside down logic; its meaning as a geological indication was plain to nobody. All sorts of geological fancies, follies and superstitions surrounded and befogged the minds of the people of every hamlet and township of the Commonwealth; and the inhabitants of the cities were still more profoundly ignorant of the whole subject.

During this year there were collected and put together, in an orderly fashion, all those ten thousand isolated facts respecting iron ore beds, coal beds, limestone ranges, salt and oil springs, lumps of lead and zinc,—respecting the red, white and neutral colored rocks which form reliable and persistent indications of where such minerals may be sought, and why money and time would be squandered in looking for them. In a word, a census of the discoveries of the past was made. An account of stock was taken of the mineral resources of the whole State, in a careful and preliminary way, preparatory to a still more minute and complete examination of each locality, and of every mineral bed of value in following years of the survey.

Of course some important generalizations were reached; for example:

In 1838 the fact first become known that the coal beds of Western Pennsylvania were deposited in two series, one above the other; that the Great Carboniferous Formation was built up in three stories. There were Lower Coal Measures, 300 or 400 feet thick; over them lay a formation from 300 to 400 feet thick, practically without coal beds; and over these Barren Measures an Upper Coal System. In other words, Mr. M'Kinney found himself, wherever he went along the Monongahela river and its branches, studying the Pittsburg coal bed and the coals of the Upper series. Mr. Hodge, on the other hand, could discover these beds nowhere in his Allegheny river district; the hill slopes being all occupied by the horizontal outcrops of the Lower Coal System. On the line of the Ohio river and the Kiskiminetas, where their two districts joined, Mr. Hodge's Lower Coal Measures could be seen sinking south-south-westward under the Barren Measures, and also the Barren Measures sinking, in the same direction, under Mr. M'Kinney's Upper Coal Series. Thus, by a happy chance, which may be said with truth never to have fallen to the good fortune of any other geologists, and which is without a known counterpart at any other place on the earth's surface, it turned out that the section which Mr. Hodge made along the valley of the Allegheny river, from the State of New York to Pittsburg, a distance of one hundred miles, exactly fitted on to and was completed by the section which Mr. M'Kinney made, up the Monongahela river from Pittsburg into the State of Virginia, a distance of another fifty miles. These two sections formed one complete and uninterrupted section, 150 miles long in a straight line, north and south, across the State. It not only exhibited every division and sub-division of the Great Coal Formation, but even every coal bed, fire-clay bed, limestone bed, iron ore bed, sandstone bed and shale deposit in those sub-divisions, from the bottom to the top; from the Red Shale of XI and the Conglomerate XII, near the New York State line, to the Brownsville and Waynesburg coal beds at the Virginia State line. Mr. Hodge even carried his section down two thousand feet below the Conglomerate, to the rocks of VIII, on the shore of Lake Erie; while Mr. M'Kinney could continue his section several hundred feet higher than the uppermost workable coal bed, to the top rocks of the country in the south-west angle of the State.

The reconnoisance which Mr. Trego and Mr. Ward made of Cambria, Somerset, Fayette and Indiana counties, revealed the same state of things in the south-eastern portion of the Bituminous Coal Region of Pennsylvania. The Lower Coal Measures were everywhere covered with a great thickness of Barren Measures, (much thicker than in Messrs. M'Kinney and Hodge's districts further west;) and also the almost complete removal of the Upper Coal Measures from the surface of the country. The remains of the Pittsburg bed, and those immediately above it in one or two isolated hill-tops were, however, not discovered, or at least certainly recognized, until the following year. But the great fact was made plain, that but one order of deposit in time and position, and but one arrangement of divisions and sub-divisions of the Coal Measures, ruled over the whole extent of Western Pennsylvania. In the words of the Chief Geologist, on page 74, of his Third Annual Report:—

"The details above presented will serve to show a remarkable regularity in the position of both the coal and the iron ores throughout indeed the whole broad belt of country embraced between the Allegheny mountain and the Chestnut ridge, and from the southern line of the State to the counties of Clearfield and Jefferson. Of the coal there would appear to be at least three seams of moderate dimensions occuring in the lower part of the series; and at a considerable elevation above these, a bed of considerable size, and usually of superior quality."

The second important generalization reached by the survey of 1838, was that of the Six Bituminous Coal Basins.

In the preceding year (1837) Whelpley and Sheafer had

worked out the general structure of the Anthracite Region; not merely fixing the limits of its principal geographical subdivisions, but showing how the crust of the earth was waved, and the beds were folded together like the sides of a canoe; canoe inside of canoe, like nests of crucibles; in a word, how the Anthracite Coal Measures, as a whole, filled up three deep troughs; each of which was, in its turn, composed of half a dozen, and in two cases, of several dozen smaller troughs lying side by side, or alternating with each other, representing several fleets of canoes moored closely together with prows and sterns interlocking.

In the succeeding year 1838 now under consideration, it was revealed that the same sort of structure characterized the Bituminous Coal region, but on a larger scale and simpler plan. Instead of a multitude of small, narrow, sharp, steep-sided troughs containing the crushed and complicated Anthracite coal beds, six long, wide, shallow, gently sloping troughs contained the nearly horizontal and hardly at all disturbed beds of the Bituminous coal system.

The first basin was seen to sweep round, in an arc of the circle two hundred miles in length from north-east to south-west, on top of and just behind, the wall of the Allegheny mountain. The Second basin passed concentrically back of the first; the third back of that; and the fourth and fifth, beyond, ever in wider sweeping arcs. Finally the sixth, issuing from the State of New York, was seen to take in all the north-western and western counties, and to carry out, at the south-west corner of the State, into Virginia the greatest pile of coal measures.

It can be easily understood that, while Mr. Trego and Mr. Ward were studying the southern portions of the first and second basins, Mr. M'Kinney was studying the southern portions of the third, fourth, fifth and sixth, and Mr. Hodge only the northern parts of the fifth and sixth. Mr. Rogers himself determined the lines of the bottoms of these troughs where they crossed the West Branch of the Susquehanna; and, probably looked at the strongly marked topography which characterizes the eastern continuations of these basins in Bradford and Tioga counties.

Such was the geological picture of Western and Northern

Pennsylvania grandly sketched out in the third year of the First Survey, 1838. The careful painting of it was reserved for after years. Although it was now known that the first basin was subdivided into two in Cambria and Somerset counties and Virginia, and that there lay in front of it, and east of the Allegheny mountain, (taken as the edge of the bituminous coal field,) other coal basins,—such as the Frostburg-Cumberland basin on the Maryland line, and the Broad Top coal basin in Huntingdon county,—it was impossible to decide by one short season's work with how much regularity these six basins kept on their course; how much irregularity might disturb the anticlinal axes which separated them; nor even, indeed, whether the basins which counted six in the north-east counties were precisely the self same basins which also counted six in the south-west counties of the State.

To decide such questions, Mr. Rogers had to wait another year. In fact many years elapsed before the many minor perturbations of this grand and apparently simple structure became known; and only now in 1874, are we really prepared to grapple with the difficult detailed geology of this distinguished portion of Pennsylvania. Simple as it looked at first, innumerable irregularities of formation are now known to exist, which require the utmost skill of the geologist to detect and estimate; irregularities, which to a certain extent vitiate the generalizations of 1838, and call for a revision, both of the numbering of the coal beds, limestones, iron ores and sand rocks, and of the geological map on which their geographical distribution is portrayed. The explorations of thirty-six years have called in question the local identity of several important beds of coal, and thrown doubts upon the identity of even such a master rock as the great conglomerate along the New York State line. Nor will our task be ended until a perfect map of the whole bituminous coal field, that is of all Western Pennsylvania, shall have been made, and every outcrop of every coal bed shall have been drawn upon it with the strictest truth and precision.

The third important result of the survey of 1838, was this: the number of anthracite coal beds had been popularly overrated.

In the previous year, 1837, Whelpley and Sheafer's study of the Pottsville basin had brought to light the important fact (on which Mr. Rogers built up his "wave theory," so famous in after years) that the north dipping beds were almost always steeper than the south dipping beds. His theory was merely an extension of this ruling fact in the small waves of the Pottsville and other anthracite coal basins, to the vastly larger waves of the mountain country of Middle Pennsylvania.

This Rule of the Steep North Dip was in 1838 brought to bear with good effect upon the identification of the coal beds. Such problems as that of the possible identity of the Lewis & Spohn beds, problems vehemently discussed at Pottsville, were settled by it. When the rule was applied to the larger basins, coal beds were identified from basin to basin. Gradually the total number of beds at Pottsville, Shamokin and Wilkesbarre (where the basins were deepest) and their order of superposition from the bottom upwards, became known. It was then clearly seen that the anthracite Coal Measures were also, like the bituminous Coal Measures, a two-storied structure; there was a Lower Coal System; there were Middle Barren Measures; and there was an Upper Coal System.

The year 1838, was therefore a year of great revelations in the history of coal geology. The identity of the anthracite and the bituminous Coal Measures was demonstrated beyond a cavil, although most, if not all, English geologists still persisted in calling the anthracite an older formation. And a long stride was made towards the actual recognition of the same individual coal beds at Pittsburg and Pottsville. When, in the winter of 1838–'39, Mr. Rogers compared the reports of his western and eastern assistants, he saw at a glance that the coal formation No. XIII had been originally deposited in unbroken sheets over nearly *all* Pennsylvania—that is, all Pennsylvania southward to a line passing through Pottsville, Huntingdon and Cumberland. And when he compared with their reports, the reports of his assistants in middle Pennsylvania, he could state with almost equal confidence that the coal area had once extended still further south, viz: as far as to Easton, Reading, Harrisburg, Chambersburg and Hagerstown; perhaps to Lancaster and York. The Pottsville, Mine Hill, Mahanoy, Shamokin, Beaver Meadow, Hazleton, Black Creek, M'Cauley's Mountain and Wyoming Valley anthracite basins in the east,—the Mehoopany, Towanda, Blossburg semi-bituminous basins in the north,—the Broad Top semi-bituminous basin at the centre, — the Frostburg (Cumberland) semi-bituminous basin in the south,—these were merely fragments of one common outspread of coal measures, now separated by erosion; and therefore outliers of the great unbroken bituminous coal field known as the Six Bituminous Basins of the west. The great synclinal troughs of Middle Pennsylvania, once as rich in coal as any in Schuylkill and Luzerne, had been swept clean of all their coal. One only, the great Huntingdon synclinal, had preserved a patch of coal measures about 80 square miles in extent, which we call the Broad Top field.

The waste between the outlying coal basins had been incalculably great. But the waste inside of the coal basins themselves had been also incalculably great. Several of the narrow anthracite basins had but the lowest bed remaining in them, or the lowest two beds. Others preserved only a part of the lower series of coal beds. Others retained these and a portion of the Barren Measures over them. While even the widest and deepest of the troughs held a residue of the upper Coal Measures also.

And the case was the same with the six bituminous troughs. In the far north-east, at Towanda, for instance, a few wafers of the two lowest beds of the lower series was all that had been spared from the wide destruction; and these lay scattered along the flat mountain tops. Further west, the lowest beds were preserved as continuous sheets, with patches of the beds above scattered over them. In the country north of the Conemaugh (Kiskiminetas) river, the lower Coal Measures occupied the entire field. In the country south of that river, in Westmoreland county, most of the upper Coal Measures also had been spared. Finally, in Green and Washington counties, the upper Coal Measures were all there, covered and protected by the upper Barren Measures.

In the preceding year, 1837, sufficient notice had been taken of the small crest-waves (anticlinal) which traverse the anthracite region, and separate the small trough-waves (synclinals) from one another,—to make it appear that they entered the large basins from one side and ranged diagonally through them. Mr. Whelpley had begun to trace and map them all. In 1838, he got the axes or central lines of these waves into their places on his map, and showed how they passed through one mountain after another; how they bent the straight mountains into hooks; and how they crossed the red shale valleys of XI, and passed through the mountains of X, into the open country of VIII.

It is not unfair to the geologists of the Old World to say that Topographical Geology was born in Pennsylvania in 1838. As there is but one such Anthracite Coal Field known in the world, so there is no field of investigation for the topographical geologist so perfectly adapted in all respects for suggesting, at sight, the principles of his branch of science, and for testing their application to all accidents of the earth's surface.

Here are no cross faults breaking the visible connection of part with part, doubling the series, and confounding the geologist. Thirty thousand feet of sediments, undisturbed by any fracture of the whole mass, are to be seen rising and falling in a few great waves, modified by scores of lesser waves, presenting to the sky the worn outcrops of every kind of rock—coal, shale, slate, sandrock, conglomerate, limestone—in ridges and valleys, of every height from a hundred to a thousand feet, and with every kind of allignment—straight, curved, hooked and scolloped. There are knobs and sugar-loafs defining the ends of the coal basins; escarpments of No. XII; terraces, marking the outcrops of coal beds; water gaps and wind gaps, occasioned by slight cross fractures, or by gentle twists of strike. There were hundreds of shafts, gangways and rock-tunnels, to give access (a little way) to the interior. There were hot contentions of boss miners, proprietors and speculators, over the underground complications; questions not to be settled by fossils, nor by mineral aspect, but solely by the geometrical study of the features of the surface. Topography was master of the situation. And Whelpley constituted himself thus the first perfect topographical geologist our science had. He received much aid from strong-minded miners at Pottsville like Richard Jones, and drew largely from the stores of local knowledge laid up by experienced coal surveyors like Samuel Miller; but he had to rely chiefly on his own genius—and what the genius of this remarkable man was there are witnesses enough to testify—for organizing the confused mass of unstudied phenomena before him into a consistent whole.

Mr. Whelpley's map of the Southern and Middle Anthracite

Coal Felds was one of the most important contributions to physical science ever made, in any country. Its eminent qualities can be appreciated only on reflecting that it was not only topographical, but geological; and that it was accomplished by himself, alone, in the infancy of geological science, with none of the modern aids, and long before the appearance of the great geological maps of Europe, none of which were at first, and indeed very few of which even now are topographical. It was substantially finished in the early spring of 1839, and engrossed in the geological map of the State in 1842. In May of 1839, its author, tired of his work, left the Survey to lead a life of singular adventure and great suffering; to be by turns a physician, an explorer, a colonizer, a writer and a mechanical inventor. He died in Boston in 1872, at the age of 55.

Mr. M'Kinley began in 1838 and completed in 1839 a topographical map of the mountain country west of the Susquehanna river, part of the middle region of the State. Bald Eagle or Muncy mountain was the northern and Shade mountain the southern limit of his field. Between them lie the Buffalo mountains, Seven mountains, Nittany, Tussey, Standing Stone and Jack's mountains; and the limestone valleys—Penn's, Brush, Nittany, Sugar and Kishacoquillas. This, also, is a region of rare geological topography. The zigzags and terraces of No. IV are its predominating feature; and the upturned-keel mountains of IV its most striking phenomenon. Its general character is, mainly, the reverse of that of the region which Whelpley studied. Here anticlinal mountains take the lead, whereas in the anthracite field synclinal mountains claim the first attention. Here M'Kinley found no Coal Measures; no red shale of XI; no terraced red and white sandstone mountains of IX and X. He was working in rocks of the Lower Devonian and Silurian systems. His mountains were all mountains of IV, in one direction (eastward) sinking slowly under broad plains of Silurian Red Shale V (with fossil ore) and Silurian limestone VI; and in the other direction (westward) abruptly terminating in fertile valleys of Siluro-Cambrian limestone II.

In 1838 the first extended study of the Clinton fossil ore of V was made. The red shale of V covers all the country between the Susquehanna river and the Buffalo mountains. It is thrown into six waves, one of which, on the east side of the river, forms Montour's Ridge, on both flanks of which, at Danville, Bloomsburg and elsewhere, the ore had already been mined. This ore was now traced westward around the end of Jack's mountain as far as Lewistown on the Juniata; through Penn's Valley; along the flanks of the Buffalo mountains; around the end and along the north side of Muncy mountain, past Williamsport to Bellefonte and Hollidaysburg.

In 1838 Dr. Jackson,—a man of singular but erratic genius, who became a distinguished physician in Western Pennsylvania, and died surgeon-in-chief of the Army of the Cumberland soon after the battle of Lookout mountain near Chattanooga, in 1863,—made a discovery of great importance, but one not appreciated for a good many years afterwards. His work that year lay chiefly in the limestone valleys adjoining Mr. M'Kinley's mountain land. After examining every iron ore bank in Nittany Valley, Brush Valley, Penn's Valley, Sinking Creek Valley and Kishacoquillas Valley, he came to the conclusion that the brown-hematite (limonite) ore of these valleys belonged to the stratified limestone beds themselves, and had been set free from them by chemical and mechanical decomposition. This view was not only new and strange to the iron masters and miners of the region, but was opposed to the prevailing feeling of the chief geologist, and not publicly accepted by him. Prof. John Frazer always resisted it on the ground of his belief that the iron had been washed out of the surrounding sand-rocks of IV, and been collected in the hollows of the limestone surface of the valleys. The manuscript reports of Mr. Bocking, twenty years later, were based upon substantially the same idea. Dr. Hitchcock, after studying the brown hematite deposits of the limestones in Vermont, Massachusetts and Eastern New York, convinced himself that all similar deposits in the limestone valleys of the Appalachian belt as far down as Alabama, had been washed into their present resting places in a very recent geological age, the Tertiary. Even in 1873 Dr. T. S. Hunt connected the easternmost range of these ores with the chemical and mechanical dissolution of the iron bearing metamorphic slates of the Blue Ridge and the Piedmont region. Dr. Newberry seems to agree with him in this; for in his excellent article "On the Iron Resources of the United States," published in the *International Review* of last November, page 764, he writes:

"A belt of limonite ore-beds passes down along the flanks of the Alleghenies from Maine to Georgia; and these deposits occur chiefly along the outcrops of the metamorphic Silurian limestones. . . . . Everything indicates that these deposits were formed by the accumulation of the oxide of iron precipitated from the ferruginous drainage of the iron bearing Allegheny highlands. In date they range from the Cretaceous to the present time," &c.

This notion, which is one of those deceptive, broad generalizations naturally and justifiably indulged in by men of science, whose knowledge is extensive, but who have not found opportunity to study the phenomena in question at sufficient length and detail in the field, was set aside by R. M. S. Jackson in 1838, after a thorough examination of the case, with eyes from the glance of which nothing escaped, and with a brain never excelled. He saw that the ore banks of Nittany Valley ranged themselves geographically into a long oval curve, along the sides and around the ends of a low barren ridge, in the middle of the valley; and that only one portion of the limestone series came to the surface *with the ores* along this oval line. Holding to this clue and following it through Morrison's cove, Kishacoquillas, and the other valleys, he found the ore banks of all of them located along the outcrops of certain special limestone beds in the series. But, had the ore been washed in from the surrounding mountain-wall of No. IV, it would have been dumped, regularly or irregularly, over the whole surface of each valley; and in equal proportions in all the valleys; for the same proportional amount of No. IV surrounds them all. When it was suggested that the vault of No. IV, which once spanned the valley, had cracked and gapped and that the iron ore had been precipitated at the bottom of the chasm, Jackson exclaimed: What a roaring torrent must have swept through that crack, swollen by perpetual rains descending mountain slopes higher than the Alps! Had he heard Dr. Hunt and Dr. Newberry conjecture that the ores lying along the eastern edge of the Shenandoah valley had been washed into it from, or from across, the Blue Ridge, he would have declaimed, in his grand dramatic way, on the absurb conclusion to which his geology of the *back* valleys would then be reduced. For what flood could suffice to sweep the Blue Ridge iron ores a hundred miles, over successive ranges of mountains and across numerous parallel valleys, in some of which, although admirably adapted to hold them, not a trace of such ores are now to be found? Mr. Rogers' flood came from the north. This must come from the south.

Had Jackson lived a geologist, he would have published a satisfactory refutation of this surface-drainage theory of the brown hematites. Special private surveys of ore tracts in the Great Valley, and in these back valleys, and especially an elaborate re-examination of part of Jackson's district of 1838 by Mr. Franklin Platt in 1873, (see his map of 100 square miles of the Nittany valley, for Lyon, Shorb & Co.) has left no ground for doubt, that Robert Jackson was the discoverer of the true argument of explanation for the existence of our brown hematite iron ores of No. II. The argument evidently made an impression on the mind of Mr. Rogers; but not one deep enough to be clearly represented in his annual report. All he says, on page 47, is this:

"Though a great number of excavations for iron ore were isited in Nittany valley, many of which are highly interesting from the richness of the ore, and the prodigality in which the mineral is supplie , and from the curious features, seen sometimes in deposits in their connection with the underlying limestone strata, yet no satisfactory explanation has been arrived at concerning the origin of the ore, nor any general practical rule discovered, calculated to conduct us with a desirable degree of certainty, in a search after large deposits of the mineral. Some light has, however, been gained from other quarters of the State, through observations made upon the same formation. If these should be corroborated by future researches, some part of the obscurity which rests upon this difficult but important subject, may perhaps be removed. For the present, I deem it premature, until the chemical examinations connected with the survey shall have been more advanced, to enter into the complicated enquiries arising out of the scientific investigation of our ores of iron."

A remarkable passage, however, occurs in a paper of characteristic temperance of judgment and beauty of style, read by Prof. William B. Rogers at the third meeting of the Association of American Geologists and Naturalists, at Boston, in 1842, which shows, that if he and his distinguished brother were prevented from freely adopting Dr. Jackson's conclusion by considerations connected with their peculiar diluvial views or from any other reason, they nevertheless held fundamental truths of sedimentary geology which could only bud and blossom into that conclusion as their native fruit.

This paper was entitled "On the connection of Thermal Springs in Virginia, with anticlinals and faults," and it gave in detail the evidence of such connection. The evolution of *nitrogen* was ascribed (suggestively) to the setting free of that gas from common air carried down from the surface to great depths by water, which, on its way thither, is heated and reascends in hot springs,—the oxygen of the air being consumed by carbon disseminated through the rocks in the form of carbonate of iron; the carbon and oxygen flying off together; the nitrogen apart, and the iron being left in the rock as hydrated peroxide, *i. e.* brown hematite iron ore. (See Trans. Ass. A. G. and N., 1842, p. 345.) The passage referred to is as follows:—

"The limestone For. II, and the slates forming a part of For. I, always contain more or less protoxide of iron and carbonaceous matter, even after long exposure to the action of the weather. When freshly taken from a new excavation at some depth, the latter rocks abound in the protoxide, and the limestone exhibits nearly all of its iron in that stage of oxidation. It would, therefore, seem probable that these and the other strata deposited beneath the Appalachian sea, contain, at great depths, this oxide to the exclusion of the sesquioxide."

This is only another mode of saying that the sesquioxide (brown hematite) deposits are the outcrop edges of the protoxide (unchanged) iron bearing layers of the Old-Age (Palæozoic) Rock System; which is precisely true, and perfectly well represents one of the chief discoveries of the third year of the First Survey of Pennsylvania, in Dr. Robert Jackson's district.

In 1838, also, it came into clear view for the first time, that the complicated geological structure of Pennsylvania differed specially from the geological structure of the disturbed Old-Age

Rock regions of Europe, in the almost total absence of downthrow faults. Now first appeared before the eyes of students of the science on this side the water, those symmetrical vaults and basins into which the seven-mile thick earth-coat of Old-Age Rocks had been pressed; to heights higher than the Himalaya mountains, and to depths deeper than the deepest ocean soundings. Arched roofs of Coal Measures, invisible to common eye, were now seen by the geologist's instructed vision spanning in air the interval between the crest of the Allegheny and the crest of Broad Top—mountains forty miles distant from each other; between the Mehoopany and Shickshinny,—between the Wyoming and Hell Kitchen,—Green and Locust,—Mahanoy and Broad mountains. In like manner the still more ancient vanished vaults of the sand-rock No. IV were restored in imagination to their places in the air over the limestone valleys of II; fragments of them still remaining, such as the bridge of stone seven miles long and five miles wide which shuts in the Nippenose Valley south of Williamsport. Before these stupendous structures, some of which must have been six or seven miles higher than the present surface, Alps and Andes hide their diminished heads.

In 1838, the erosion of the earth's surface was first clearly revealed to geologists in America in all its magnitude, and took its place as prime factor in their calculations. It was not adequately felt by European geologists until long afterwards. M. Desor, although born in the Jura, and bred in the Alps, expressed unbounded astonishment and delight at the first fair sight he got of the erosion of our anticlinal valleys in 1851. The late lamented chief of the Irish geological survey, Prof. J. Beete Jukes was the first English geologist who saw it himself and taught it to others. The doctrine of vast subaerial erosion, which now plays so prominent a rôle in all discussions of the physics of earth, and especially of dynamic and sedimentary geology, may therefore be justly called a Pennsylvanian discovery, dating from the years 1837 and 1838.

In 1837, Whelpley discovered that existing mountains are the remains of a continent once standing at some easily determinable higher level than the present continent. He saw that this must be true, as soon as he attempted to map the synclinal plateaux of the Anthracite Coal Field. And in 1838, this master-key thought in structural geology became the common property of the Assistants on the Pennsylvania Survey. Since then it has not only supplanted, but sent out of existence, the old time notions of "mountains of elevation," and has set in a true light the origin of the newer and loftier ranges of the world. It affords us ampler materials out of which to get material enough for the construction of the low countries; and it explains the production of the later rock-systems out of the debris of the earlier, through all geological times.

Unfortunately Mr. Rogers was led by an overstrained poetic sentiment to imagine for these tremendous phenomena a physical cause which no one now considers to be necessary for their explanation. But it must be remembered that no one questioned *then* that the interior of the earth was a globe of liquid fire, upon which a cooled crust floated. Mr. Rogers used the accepted theory of earthquakes. Earthquakes of any required magnitude and violence might have happened. He had only to imagine earthquake waves, in the fiery ocean underneath, sufficiently large to permanently fix a crust say ten miles thick, into these immense vaults and cradles, and, if the lava waves moved north-westward, the steeper dips in that direction would be accounted for.

The same poetic temperament and large way of looking at things induced him to use the then current belief in ancient deluges for explaining to himself and to his assistants the sweeping away of the upper formations, and the removal of the rock-wave crests. He thought he could bring his moving ocean from the north, and imagined it to have been generated by the rise of the continent at the close of the Coal Era.

Such poetic dreams have passed away. Such mere theories have been covered up and fossilized beneath the slow accumulations of common sense investigation in all the fields of natural science. They served their turn to inspire with enthusiasm the first workers; but so far as they influenced the minds of the Assistants, they did harm in this one respect, namely: they prevented a quite intelligent study of the actual surface. For, every patch of boulders, every gravel bank, every isolated knob in a gap, every scratch upon the bare rocks, and even the sub soil loams and sands, were referred, more or less, under this theory, to some other locality further north. Had it happened *after* Agassiz had studied ice, instead of before,—had Mr. Rogers himself seen the glaciers of the Alps,—his was just the genius to anticipate the discovery of the "glacial theory," and his fame would have been even greater than it is. It was not until Desor joined the Survey, in 1851, that we learned how important a part of our geology the subsoil is; how much part and parcel of the rocks beneath it, it is; and therefore, what a guide it is in exploration.

The fourth year of the First Geological Survey was 1839.

The annual report of the Chief Geologist, sent in by the Secretary of the Commonwealth, February 8, 1840, makes a printed pamphlet of 215 pages, 45 of which contain analyses of iron ores by Dr. R. E. Rogers and Mr. Martin Boyé, chemists of the survey.

The field work of this year was substantially a continuation of that of the preceding year; most of the Assistants pursuing the study of their respective belts of country towards the south and west.

The cost of the survey in 1839, amounted to $16,000, (15,-991 27.)

Mr. Whelpley and Mr. Sheafer resigned from it; and Dr. Andrew A. Henderson of Huntingdon, and Mr. Peter Lesley, Jr., of Philadelphia, were added to the corps. Mr. Stone served as a volunteer throughout the field-work season.

Mr. Holl explored the south-east corner of the State, south of the New Red Formation, viz: The Bristol and Germantown hills, Chester, Delaware and (south) Lancaster counties. *First District.*

Mr. Trego was withdrawn from Somerset and Fayette to explore York and Adams counties, which had been neglected in 1838; and also the Cumberland Valley. *First and Third District.*

Mr. Whelpley remained at Pottsville until June to finish his anthracite map, and to induct Dr. Henderson and Mr. Lesley into the geology of the Coal Region. Dr. Henderson, however, was soon assigned (in May) a separate district west of the Susquehanna river, and Mr. Lesley remained to continue alone the collection of information furnished by the collieries, and to enlarge the borders of Mr. Whelpley's map, by taking in Devonian country to the north and south of it. *Second District.*

Dr. Henderson commenced a systematic survey of the belt of country lying between Shade, Black Log and Cove mountains on the south, and Jack's mountain and Sideling Hill on the north. *Third District.*

Mr. M'Kinley continued the study of his belt of the Seven Mountains through Stone Valley, across the Juniata river, to Broad Top; keeping Jack's mountain and Sideling Hill on the south, and Tussey mountain on the north. *Third District.*

Dr. Jackson continued the study of his belt, between Tussey mountain and the Allegheny mountain, as far south-west as Bedford. *Third District.*

Mr. M'Kinney, assisted by Mr. Ward, undertook to measure *instrumentally* a section across Westmoreland and Washington counties, from Mt. Pleasant at the foot of Chestnut Ridge to the Ohio River at Wellsboro'. His lines passed along the turn pike; Big Sewickley creek; across to Williamsport; down the Monongahela; up Mingo creek and down Cross creek to the Ohio river. This line was regularly surveyed; and the heights of the coal outcrops were determined with the boiling-point thermometer and barometer.* They made also exploring journeys through what is now known as the Oil Region, and through Mercer and Butler counties, in the field occupied in the previous year by Mr. Hodge. *Fifth District.*

Mr. Hodge, accompanied by Mr. Stone as a volunteer, and provided with a pack horse, tent and two men, commenced a systematic exploration of the First and Second Bituminous Coal Basins lying on the high Allegheny Mountain table-land of Wyoming, Sullivan, Lycoming and Clinton counties, cut through by the deep gorges of the Loyalsock, Lycoming, Pine, Larry's and Tangascootac creeks, and by the North and West Branches of the Susquehanna river. All old coal openings in the woods were visited, and new ones were made. All the

*The use of the boiling-point thermometer as a barometer in the First Survey was not successful. The instruments were all broken in the carriage, and no records of observations were ever published. This line across the Western Coal Fields does not appear in the Final Report.

patches of Coal Measures left on the high table mountains, as far north as the New York State line, were distinguished from each other, and located upon the old map of the State; and a very clever approximation was made to a knowledge of the number, thickness and quality of the coal beds in them.

Mr. Rogers now saw his way clear to hazard a promise in his annual report, at the end of the year 1839, that one year more should "complete nearly the whole of the field explorations, leaving but few tracts in the State, of any geological interest, unexplored." A part of still another season, however, would be required for revising intricate portions of the geology, clues to which had been found in neighboring districts; for certain tedious measurements of high practical importance; for getting into shape the general geological map of the State, and sundry local maps and sections; and for completing the chemical analysis of the large collection of ores, coals, cements, fire-clays, &c., which were intended to have written analyses attached to them, when arranged in the future museum at Harrisburg.

In 1839 there were sent to Philadelphia 130 boxes, containing about 3,000 specimens. The whole collection had reached 8,000, of which 4,000 had been temporarily classified for analysis and reference.

The gains of the survey in 1839 may be summed up in one sentence: The new geological views reached in 1837 and 1838 had been this year applied to and been found to hold good for all parts of the State; while vast accumulations of local facts had been made for publication in a final report.

In the south-easternmost district, extending from Trenton to Port Deposit a very good advance was this year made in tracing the beds of gneiss, mica slate and serpentine. Mr. Rogers found himself strong enough in facts to pronounce boldly against the igneous origin of serpentine. This was an auspicious omen for our emancipation from a whole set of geological superstitions. Years passed, however, before the conviction could be entertained that beds of magnetic iron ore were also sedimentary, and not cooled and crystallized iron lavas.

Numerous streaks of marble or crystalline limestone were found running through the Chester county country. Beds of chromic iron and titanic iron ore were brought into notice. Localities yielding zircons, beryls and large garnets, agates, tourmaline, corundum, sapphire, asbestos and porcelain clay (kaolin) were visited and described; the margins of gneissl and, mica-slate land and talc-slate land, with steatite (soap stone) and serpentine, were defined with some truthfulness upon the map, and the Peach Bottom roofing slate of the Susquehanna was studied.

But after all, no good account could be given of the true structure of this most interesting district; nor had any close approach been made to an understanding of the age of this belt of rocks. And in the same dark state it has lain ever since. The fact was, the First Survey had its hands too full of other more pressing matters, and was too poor, to make a *thorough instrumental* survey in this or in any other district; and that being the fact, no accurate knowledge could be got of a land ot uncommon complexity of structure, without fossils, and under cover of deep decomposed subsoil, by merely walking over the ground, visiting quarries and rocky hill sides, and collecting specimens.

The same was true of the ancient rock country north of the Chester Valley.

Even the structure of the Chester Valley itself was left uncertain; although every marble quarry and iron ore bed in it, from Willow Grove at its eastern end, past Downingtown, to its western end in Lancaster county was visited, and the well exposed rocks along the Schuylkill river banks were examined, measured and sketched over and over again.

The geology of York and Adams counties, so far as the oldest rocks were concerned, was also left at the end of the year in a very unsatisfactory state, little being known of it (and less still of the South Mountain range) except that all its rocks were older than those which occupied the rest of the State; and that such and such minerals could be found in them.

The year 1839 was remarkable for Pennsylvania geology by the experiments of Mr. William Lyman of Boston, at Pottsville, to smelt iron ores with anthracite coal; and also by the arrival of Mr David Thomas and his family from Wales, to attempt the same feat near Allentown. Mr. Lyman, after many months of anxiety, at first alone, and afterwards assisted by the experience of Mr. Thomas, was successful; but his exposures and labors by day and night undermined his constitution, and he died a few years afterwards in New England. His name should be held in the front rank of honor, as one of the pioneers of that industry which has made Pennsylvania in a special manner famous. Mr. Thomas had been of the beginnings of the industry in South Wales, and was sent over by Mr. Crane, the rich iron master of England, to establish it on the Lehigh; built the first Crane furnace in 1840; afterwards built furnaces of his own; and amassed a fortune well deserved and well employed. Hale and active at eighty years of age, he sat as patriarch, chieftain and sage among the assembled iron men of the United States, at the banquet given in Philadelphia, on the 10th of December 1874, by the American Iron and Steel Association to its English guests, Mr. I. Lowthian Bell, president of the Iron and Steel Association of Great Britain, and Mr. Thomas Whitwell.

The Pioneer Anthracite Steam Furnace at Pottsville was built in 1838 expressly for repeating the Welsh experiment with Pottsville anthracite. Tradition says that, 12 years earlier, in 1826, a little furnace, one mile below Mauch Chunk, made iron successfully with Lehigh anthracite. Tradition of a more specifie and reliable kind shows that anthracite-burning locomotives habitually drew trains of coal cars up the steep gradients of the Beaver Meadow railroad for a number of years before such a thing was thought of in any other part of the world.

The Chief Geologist was now besieged with solicitations to discover black band, or other iron ore deposits, in the anthracite Coal Measures; and he gave directions that every outcrop of iron bearing sandstone or shale should be examined, and all possible care should be taken to detect the presence of the much coveted mineral. But the best of geological surveys cannot discover what does not exist, nor make available for practical use what costs too much to procure.

Already before the close of 1839 it became tolerably clear that no extensive iron mines would ever be wrought within the limits of the Anthracite coal fields proper; and the history of 35 years,—with its record of gangways driven and shafts sunk to follow or to strike beds of coal-measure ironstone and blackband ore,—confirms the soundness of the geological sentiment on this subject then formed. There may exist exceptionally thick and extensive sheets of ironstone somewhere underground, concealed in the measures which lie between the coal beds; and such a plate may be accidentally struck in any month of any year hereafter; who can tell? But until such an accident shall occur, the idea of the essential barrenness of the Anthracite rocks in iron ore, got by the studies of the Survey in 1839, will be retained. Such an accident, however, is not without example. In 1874, a nine foot bed of ironstone was found by an English engineer underlying the soil of a valley in the north-eastern district of England, close to Mr. Bell's iron works, and that at a depth of only eight feet. It was wholly unsuspected, although old and extensive mines of ironstone overlooked the locality from the hills above.

The year 1839 may be called the Iron Year of the First Survey for yet other reasons. Mr. Hodge, in his systematic study of the coal patches thinly coating the Allegheny table land, had to lead his camping party innumerable times over the vertical outcrop-escarpment of the Conglomerate No. XII, which makes a cornice around all the mountain spurs of that country. Underneath this hanging wall of horizontal rock, (the home of millions of rattlesnakes,) crowning slopes which rise about a thousand feet in height from the main streams, he found the soft red shales of XI; and in their upper part, close to the base of the Conglomerate, the dove colored carbonate iron ore of XI; the first of that series of iron ore deposits which belong to the Coal Measure Age. With this exhibition he was so profoundly impressed, especially wherever it showed itself at the best, (for example, at Ralston, in the Lycoming valley) that in after years he decided to invest the savings of his laborious life in it, in preference to the Lake Superior copper,—and lost all he had. Mr. Rogers, and those of his other assistants who had districts containing such rocks, were all alike impressed with the conviction that this bed of iron ore, which was sometimes four feet thick, and seemed to lie in an almost unbroken sheet beneath one-half the State of Pennsylvania, would be one of the chief resources of our Iron Industry. Perhaps ere this it would have become so but for the discoveries of richer ores, in greater masses, and of easier access, in Michigan and Missouri. Perhaps it will yet become so when population condenses, wages fall, and capital offers itself cheap for investment. But thus far the value attached to this deposit by the Survey of 1839 has seemed extravagant.

The tracing of this iron ore bed by Mr. Hodge through so large a district of Northern Pennsylvania this year (1839) afforded the required geological explanation of the sub-conglomerate kidney iron ore deposit, with coal, which he had reported upon the year before (1838) at Meadville and other places in the far north-west; and to smelt which a small quarter-stack blast furnace was built in 1838, with arrangements then novel, for taking the gases from the tunnel-head to raise steam for the blast engine.

In 1839, during the careful study of all the layers of the Lower Bituminous or Allegheny River Coal System, the first general report was made of the "Buhr-stone Iron Ore" deposit of Clarion and Butler counties; on which Mill Creek, Webster, Beaver, Shippenville, Lucinda, Madison, Clarion, Etna and Jackson furnaces had been built at various dates between 1832 and 1839. This brown-hematite deposit was as important to the Charcoal Iron Trade of America then, as the brown hematites of Eastern and Middle Pennsylvania are to the Anthracite Blast Furnace production of the country now; being a deposit co-extensive with the Lower Coal Measures in Pennsylvania, Ohio and Kentucky, and already in 1839, serving eighteen furnaces in South-Eastern Ohio, viz: Brush Creek, built in 1812, Old Steam and Marble (1816,) Union and Franklin (1826,) Pinegrove (1828,) Sciota (1830,) Bloom, Clinton and Etna (1832,) Lawrence and Vesuvius (1834,) Hecla and Mt. Vernon (1835,) Lagrange and Buckhorn (1836,) Jackson (1837) and Centre in 1838;—and at least eight furnaces in Eastern Kentucky, viz: Steam (1817,) Argolite (1818,) Bellefonte and Red River (1828,) Amanda and Raccoon (1831,) Caroline and Clinton, built in 1833.

Mr. Rogers on page 163 of his annual report of 1839 writes of this remarkable "Buhr-stone" iron ore deposit, thus:

"The great importance of this valuable deposit to the growing wealth and enterprise of the western part of the State, is beginning to be appreciated. I need hardly, therefore, dwell upon it any further than to mention the following circumstances in its favor: The richness of this ore, far exceeding that of the nodular variety common in the slates of the coal measures, will bear comparison with much of the best ore found in the limestone valleys south-east of the Allegheny mountain. Another merit is the facility with which it may be smelted either alone or mixed with the nodular or ball ore, and particularly with the bog ore found so abundantly in Venango and Clarion counties. In the thicker parts of the deposit it is a nearly pure semi-crystalline protocarbonate of iron, a variety well known to be susceptible of extremely easy reduction. It is moreover recommended by the readiness with which it can be traced, arising from its accompanying closely the fossiliferous limestone, which furnishes an excellent landmark. The buhr-stone assists the discovery of it in a still greater degree, the fragments of that peculiar stone being so easily recognized, serving to point out the position of the ore. Besides these important facts, it should be mentioned that a workable seam of coal occurs above the ore and limestone, and another beneath them at a moderate interval. The importance of this triple association of the materials employed in the manufacture of iron, to the future wealth of our western counties must be obvious to every one who adverts to the incalculable advantage which Great Britain has derived from the same fortunate union."

Nor could the merits of the Buhr-stone ore bed well be exaggerated, so long as the forest timber, which overgrew the hills above it, held out for the supply of charcoal. When the "Iron Manufacturer's Guide to the furnaces, forges and rolling mills of the United States" was published in 1859, its tables gave a list of 104 charcoal and coke blast-furnaces in North-Western Pennsylvania; 17 in North-Eastern Ohio; 48 in South-Eastern Ohio and 24 in Eastern Kentucky—193 in all; most of which had once obtained, or were then obtaining their stock chiefly from it. The combined yield of 128 of these stacks in blast in 1856 amounted to 170,000 tons of charcoal metal. A few furnace men had ventured successfully on the use of coke or raw coal. Now that the production of strong pure coke from crushed and washed bituminous coal is become a manufacture, we may expect that the area occupied by the Buhr-stone ore deposit will be one of the important iron-making regions of America, and at no very distant day.

In 1839 Mr. Rogers published the first chemical memoir on the iron ores of Pennsylvania, in the form of a connected list of analyses made in the laboratory of the Survey. It occupies 45 pages of his annual report.

In 1839 the Larry's Creek "fossil ore" lying at the base of No. IX (Catskill) was brought to the notice of the Survey by Mr. Hodge. It is now known to outcrop irregularly along the whole extent of face of the Allegheny mountain. Mr. Cessna, of Bedford, has recently had it opened, for a length of ten miles, back of the Huntingdon, Broad Top, Bedford and Cumberland railroad; and analyses by Mr. M'Creath, of specimens sent to the laboratory of the Survey by Mr. Homer Young, from Milanville, in Wayne county, near the Delaware river, will appear in this year's annual report. This ore singularly resembles the much older deposit at the bottom of No. V, (Clinton) and is probably as extensive; underlying the whole Bituminous Coal Field at a great depth.*

In 1839 the first systematic study of a far more important iron ore deposit, at the bottom of No. VIII, was made by Dr. Henderson, who found it pretty extensively mined along the Juniata river above Lewistown; as it was, in Mr. M'Kinley's district, on the banks of Yellow creek in Woodcock valley. Dr. Henderson's observations of it led him shrewdly to an important discovery in a geological sense, although his conclusions diminish largely the future practical value of the deposit. The ore thus mined along the outcrops of the bottom slates of VIII (just over the Oriskany sandstone, VII) is a soft brown hema-

*In October, 1875, this ore was found in the same transition layers VIII-IX, (top of Chemung and bottom of Catskill,) in Aughwick valley, in a cutting of the East Broad Top railroad. It is represented along the northern line of the State by one of the "Mansfield ore beds" of Tioga and Bradford counties. In Western New York it is represented by highly ferruginous red layers on the Genesee. See James Hall's report of 1843, page 278, and J. F. Carll's Report of Progress for 1874, pages I. 99 and I. 100.

tite, and continues to be of that nature down to a certain variable depth, say 100 feet beneath the surface. Dr. Henderson saw that this was not the original condition of the ore formation; that this character was a product of decomposition, chemical and mechanical; that the bed would always be found to yield to deep mining some other kind of ore. In the brown-hematite of some he found lumps of impure carbonate of iron, and he believed that the entire ore bed below a certain level was an earthy ironstone. This level would be the drainage level of the country.

Dr. Henderson's anticipations have been partially realized. In late years huge masses of "rock ore" have been met with; and, as Mr. Dewees' report will show, tunnels driven across the measures to get a deeper hold of the brown-hematite have struck the bed where there was no brown-hematite at all; the whole bed being a stratum of solid undecomposed earthy ironstone, like those which occur abundantly in the Coal Measures. In fact the deposit of ironstone just overlying the Conglomerate No. XII is a repetition, in a subsequent age, of this deposit of ironstone just overlying the Oriskany sandstone, No. VII. And it is not at all unlikely that this again is merely a repetition at the beginning of the Devonian age, of deposits of iron in the form of carbonate, now wholly changed to oxide, which took place, at the beginning of the Siluro-Cambrian age, on top of the Potsdam sandstone, No. 1. The different degrees in which the carbonate has been turned into oxide (clay ironstone into brown hematite) would then be proportionate to the respective and vastly remote ages of the three formations; the first marking the beginning, the second the middle, and the third the end of the great Old-Age (Palæozoic) system.

In 1839 a flood of light thus began to pour in upon the before almost unknown Geology of the Iron Ores. The discovery of Jackson in 1838, was supplemented by the discoveries of Henderson in 1839; and both were supported and illustrated by the independent observations of Mr. Trego along the north flank of the South mountains between Harrisburg and Hagerstown.

For, at the close of 1839, Mr. Trego visited all the ore banks of the Cumberland Valley; and although he constructed no theory, he recorded many facts, which were summarily published in Mr. Rogers' annual report for 1839, and which, when viewed in the light of our present knowledge, suffice to overthrow the drift-theory of the brown hematite beds, and to confirm Jackson's and Henderson's views.

Mr. Trego noticed with much curiosity the important fact, that two separate belts of different kinds of brown hematite ore beds follow the foot of the South mountains: one of cold-short ores next to the Potsdam, and another of red-short ores in the limestone. In Prof. Prime's preliminary report of his Lehigh district the first important step made towards the true statement of the cause of this difference will be described. Everything on this subject has been vague enough hitherto. The silicon in the sandstone and sulphur in the limestone, have had the credit of making this distinction between the cold and red short hematites. Perhaps they deserve it; but it is becoming more apparent every year, that the ore is never found *in* limestones, but *between* them; that it has not come from the decomposition of limestone (dolomite) layers, but from the decomposition of sandy and slaty layers lying above, between or underneath the dolomites.

Prof. Prime will show, in his report, that one range of brown hematite ore beds west of the Lehigh, are connected with *potash slate* (damourite) beds. The ore seems, so far as his observations as yet extend, to have been set free from the weathered outcrops of these potash slates, which have been noticed by other observers in other parts of the State. But Prof. Prime is the first geologist who has assigned their probable importance to these peculiar slates of the Siluro-Cambrian formation.

In 1839 also the first systematic study was made of the much older (Cambrian) iron ores of York and Adams counties, by Mr. Trego, but it was of the most irregular and superficial character, and amounted to little more than a reconnoisance of the ground. Two short continuous belts of ore were recognized by him; one, along the south edge, and the other along the north edge of the narrow zones of Siluro-Cambrian limestone No. II, in which flows Codorus creek, and which stretches from the Lancaster county limestone country into Maryland;

the first one running just south of Hanover and Littlestown, and containing much manganese; the other skirting the south foot of the Pigeon Hills. No subsequent extensive survey of this ore region was ever made; but it has occupied public attention greatly of late years; and much capital has been invested in the development of its mineral resources without additional light being thrown upon the age and structure of its rocks.

Prof. Persifor Frazer, Jr. will state in his report the results of the instrumental survey of 1874; and his map will show that six or seven belts of ore range across York county into Adams; some of them in the Cambrian rocks, others in the New Red. This is but preliminary to a more complete investigation in 1875. Such a question as that of the true structure of the oldest and most complicated system of rocks in our State, and a system holding iron ore beds, ought no longer to be left open to barren conjecture. If the skill of modern geology is of any avail where difficulties accumulate, it should be expended here; and we may cheerfully hope for good results. Already it suggested itself to Mr. Roger's mind that the two belts of iron ore, bordering on both sides the York county limestone, represent the belt of ore bordering the limestone of the Carlisle and Chambersburg Valley. And if so, then the neighboring rocks are Potsdam No. I, under an altered aspect. He calls the Pigeon Hills No. I, (Potsdam.) We cannot connect our geology with that of Western New England and Canada unless we be permitted to speak with confidence of the presence in Pennsylvania of the "Quebec Group." The two districts of York and Adams, and of Chester, Delaware and South Lancaster, are the fields on which this important geological problem can best find its solution. The fault along (on the south side of) the Chester County Valley makes the study of it there very difficult. We hope to grapple with it more successfully along the Codorus creek.

The Fifth year of the First Geological Survey of Pennsylvania was 1840.

The work of this year was an enlargement and development of that of the preceding; and the expenses of the survey reached their maximum of $17,800. The annual report of the Chief Geologist, bearing date February 1, 1841, accounted for seven months of continuous field work, extending from the middle of April to the middle of November. The assistants in charge of districts were now in excellent train, and experienced in research, so that the scientific harvest of the season was more abundant than in any previous season ; while fewer discoveries of a notable character were made ; in fact, because such discoveries were no longer possible. The science of Pennsylvania geology had now passed safely through its arduous terms of infancy and childhood, and had attained a healthy and vigorous maturity. The chief geologist was not only in possession of an eminent point of view from which to see clearly all the field, and to direct the most efficient application of the forces at his command, but all uncertainty of action was at an end, and facts fell into system as fast as they were collected.

During this and the previous years, in company with his brother W. B. Rogers, the Chief Geologist of Virginia, he had traveled widely over the United States. What he thus learned abroad reacted wholesomely upon the survey at home. The geological structure of the Appalachian formations were thus studiously followed from Virginia to Alabama, and from Pennsylvania to the banks of the Hudson.

In the meanwhile the New York geologists, whose surveys also commenced in 1836, had put in order all the northern outcrops, from the mountains of Massachusetts to Upper Canada. The United States geologists had made discoveries in the region of the upper lakes. The outline form, as well as the solid contents of the Old-Age ocean, which once covered the United States, and in which the Palæozoic formations from Potsdam to Coal Measures had been deposited, was by this time (1840) pretty well known. The limits of the coal fields were fixed. The system of great downthrow faults in the South had been observed, and the gradual passage into it of the northern system of great anticlinals. The extension of the fossil ore* beyond

* Not, however, as then thought, the *upper* Clinton Fossil ore beds of Danville and Frankstown in Pennsylvania, but the *lower* or *rock fossil* ore bed at the base of the Clinton, mined by R. H. Powell, on the flank of Tussey mountain south of Huntingdon. The Pennsylvania fossil ore proper seems to be wanting in Tennessee.

Tennessee had been determined, and some idea had been got of the most ancient continental rocks, which formed the mountainous shores around the Palæozoic sea.

In 1840, therefore, the geologists of America were prepared to come together in consultation. On the 2d of April, they met in the rooms of the Franklin Institute, and organized themselves into an "Association of American Geologists." Dr. Edward Hitchcock, State Geologist of Massachusetts, was elected chairman, and with him were present the geologist of Vermont, Mr. C. B. Hayden, the New York geologists, Lewis C. Beck, Lardner Vanuxem, Wm. W. Mather, Ebenezer Emmons, James Hall and Timothy A. Conrad; the Pennsylvania geologists, Henry D. Rogers, Charles B. Trego, Martin H. Boyé, Robert E. Rogers and Alexander M'Kinley; the geoligist of Delaware, James C. Booth; the geologists of Michigan, Douglass Houghton and Bela Hubbard; the distinguished English geologist, Richard Cowling Taylor, and the eminent geologist employed by the United States government, Walter R. Johnson, both of these last mentioned being residents of Philadelphia.

The influence of this association on the growth and spread of science in America cannot be overestimated. The second of its meetings took place during the second week of April, 1841, in the rooms of the Academy of Natural Sciences in Philadelphia, when Dr. Hitchcock delivered his retiring address, and Prof. Silliman was elected to preside. At its third meeting, (1842,) held in Boston, Prof. Silliman read his address, which was long remembered and talked of, and Prof. H. D. Rogers was elected chairman. At the fourth meeting at Albany, (1843,) Prof. Rogers delivered his address, in which he resumed his own and his brother's knowledge of Appalachian geology, and announced their new nomenclature. The fifth meeting was held at Washington, in May, 1844; the sixth at New Haven, in April, 1845; the seventh in 1846, and the eighth at Boston, in September, 1847.

At each meeting new members were received, papers were read, important discoveries were announced, and individual views were well argued *pro* and *con.* A few short, and now all too precious pamphlets, containing the minutes of the meetings, and revised addresses, were published, and this slight literature is all that visibly remains of these annual displays of intellectual activity the effects of which upon the assembled field workers were deep and wide in their bearing upon the theory and practice of geology in America; re-acting on all public and private surveying; counteracting charlatanism and ignorance, and reinforcing the spirit of every genuine investigation. A large proportion of the men who were systematically pursuing any branch of natural science attached themselves, one by one, to the society, which broadened its title to receive them, soon calling itself the "Association of American Geologists and Naturalists."

By the time, however, that its eighth meeting in September 1847 arrived, working geology had received a death-blow. Most of the State Surveys were in the popular phrase "finished." The financial panic and crisis of that year stopped speculative investments. Mr. Agassiz, after revolutionizing some of the most important geological views of Englishmen, had settled in America. That meeting was held in Boston, where natural history was weak and physics strong. The physicists of Cambridge demanded entrance and, being admitted, seized at once the government of the old association, and changed its aspects and its polity. Its very name was ignored; the numbering of its annual meetings was stopped; and its memory was swallowed up in the "Proceedings of the First Meeting of the American Association for the Advancement of Science, held at Philadelphia, September, 1848."

This episode is needful to explain the state of things under which field work in Pennsylvania was recommenced in the spring of 1840, and continued to its close.

Mr. Rogers himself revised the *first district*, lying south and east of the South mountains, for his final report.

Mr. Trego crossed to the east side of the Susquehanna river and made a general survey of the Great Valley through Dauphin, Lebanon, Berks, Lehigh and Northampton counties; sending specimens from the ore mines recently opened along the foot of the Reading and Allentown hills, to Mr. Boyé, to be analyzed for the benefit of the Anthracite Iron Industry iust establishing itself upon the Lehigh river.

Mr. Trego was then transferred to the Devonian country between the Anthracite coal fields and the Susquehanna river, to gather topographical data for a geological map of the State.

Mr. Lesley made a topographical and geological sketch map of the Devonian and Silurian belt between the Kittatinny mountain on the south, and the Pottsville and Mauch Chunk coal field on the north, extending from near Harrisburg to the New York State line at the Delaware river. The Port Clinton group of anticlinals were arranged; and the geology of the Pocono moun tain at Lehigh, was correlated with that of the Catskill mountains in New York.

Mr. Lesley then joined the camp of Mr. Hodge, at the summit of the Allegheny, and explored with him the Somerset and Fayette county coal basins.

Dr. Henderson perfected his topographical and geological sketch map of the belt of closely complicated hills and vales ranging through Perry, Juniata, Union, Huntingdon and Franklin counties, from the Susquehanna river to the Maryland State line. This beautiful product of a rare genius for patient, close and accurate investigation, is one of the monuments of the first survey. As an illustration of geological science, it leaves little to desire, and as it was never separately published, but engrossed in the general geological map of Pennsylvania, it deserves publication still. Mr. Rogers always spoke of it with admiration and does not conceal this sentiment in his annual report.

Mr. M'Kinley finished his study of the middle belt, including the Broad Top coal field, from Huntingdon to the Maryland State line. But the complicated structure of Broad Top completely foiled him, and no wonder; for the writer some years later spent two entire seasons there, setting more than 11,000 stakes on run and leveled lines within its limits, without entirely unraveling the plot of movement in its Coal Measures. Mr. Fulton has been at work upon them ever since; and still there is much to learn. A few months of additional instrumental work, however, will finish an elaborate topographical and geological map of this most interesting and important little semi-bituminous coal field.

Dr. Jackson continued and completed his study of the Bedford-Somerset belt facing the Allegheny mountain as far as to the Maryland line, and collected materials for the State map; although he had no skill for map-making himself, and did not attempt it.

Mr. M'Kinley and Dr. Jackson filled up the residue of this season's work in the south-west corner of the State.

Mr. Hodge carried southward his survey of the first bituminous coal basin from where he left off in 1839 upon the Tangascootac in Centre county, through the Beech Creek, Snow Shoe, Clearfield and Mt. Pleasant country. Mr. Ward was his companion this year instead of Mr. Stone, and his camp consisted of two tents, a wagon and three horses, with three brothers for his working hands, two of whom he had trained the previous year, and a cook, who hunted and could occasionally dig. In August, Mr. Lesley joined this party, which had by that time reached Loretto. The camp was then moved from station to station every week or ten days, down the Conemaugh and up the Sandy, down Castleman's river, and back through Ligonier valley, and was broken up when a foot of snow was on the ground.

Mr. Lehman, the talented artist of the Survey, was in this camp for more than a month. The rest of the season he was sketching elsewhere in the State. Mr. Hodge confined his explorations chiefly to the iron ore beds, and Mr. Lesley his, as much as possible, to the coal beds. The ore of XI was studied along the face of the Allegheny mountain, and on the anticlinal mountains bounding Ligonier valley. Every effort was made to comprehend the character, order and extent of the ore deposits of the Coal Measures. The Cambria Iron Works at Johnstown, the largest in America, have made these deposits famous. The proprietors have recently employed Mr. John Fulton to make an instrumental geological survey of them, and granted leave to the Board of Commissioners of the State Survey to publish Mr. Fulton's map and report next winter.

Dr. Jackson and Mr. M'Kinley, after finishing their belts of country in Middle Pennsylvania, spent the last three months of the season in Greene and southern Washington counties, tracing the outcrops and making rudely measured sections of the Upper Barren Measures overlying the Waynesburg seam, to make sure that they held no workable bed of bituminous coal, and to complete our knowledge of the coal series of Pennsylvania to the topmost layer.

Mr. Holl was occupied this year in Mr. Hodge's old district of Erie, Crawford, Mercer, Butler and western Venango. A perfectly satisfactory account had not yet been given of the very difficult geology of the lowest part of the Coal Measures bordering on the Lake Erie high lands; nor was enough yet known of the way in which the Barren Measures influenced, for practical purposes, the shape of the ground between the Allegheny and the Beaver rivers, where they covered the workable beds of the Freeport system. All this was pretty well made out by Mr. Holl in a general way. But no part of Pennsylvania calls now for an accurate instrumental survey more imperatively than this, which has become the great oil producing region.

The most striking discoveries of the Survey in 1840 were: the discovery of the great fault of the M'Connellsburg Cove in Franklin (now Fulton) county, by Dr. Henderson; the detection of the small outlying fragment of the Pittsburg coal bed in the Ligonier valley; and the certain identification of the same bed with others above it at Salisbury in Somerset county, by Mr. Hodge's party. Mr. M'Kinley had noticed another minute fragment of it left at the summit of the Broad Top mountain, but does not seem to have recognized its identity; nor is any allusion made in the Final Report to the immense *theoretical* importance of this occurrence.

But the really important fruits of the Survey in 1840 were of a nature not to be specially stated: a completion of the preparation for publishing a true map of Middle Pennsylvania, one of the most complicated and curious portions of the earth's surface; and a definite establishment of the order of the coal beds over all Western and Northern Pennsylvania, leaving no room for doubt that one law governed the whole deposit, and that it could be safely asserted that about a dozen workable coal beds could be identified everywhere within the limits of its eroded edges. It was also made perfectly evident, that the interval between any two beds perpetually varied within certain limits; and that the thickness and quality of each bed varied incessantly, for any one who wished to trace it from one point in its area to another. This became still more evident after Dr. Jackson's study of the Westmoreland coals in the following year, and the evidence has gone on accumulating in the hands of all the western geologists to the present day. The last expression of it will be presented with this First Annual Report of the Second Survey of Pennsylvania, in the shape of a memoir on the subject, by Prof. J. J. Stevenson: "On the alleged Parallelism of Coal Beds."

The Annual Report for 1840, made a pamphlet of 179 pages, 68 of which were taken up by suites of analyses of ores of iron and zinc-lead, cast iron, coals and limestones. The rest of the report was given to a description of the Reading and Easton range of the South mountain, and to the description of the First, Second and Third Bituminous Coal basins between the West Branch of the Susquehanna and the Maryland State line.

The publication of the results of the Survey here ceased; for the Annual Report of 1841 was a small pamphlet, merely stating in general terms the distribution of work during that year. A vast mass of materials was reserved for the Final Report, which, however, did not appear for seventeen years, viz: in 1858.

Mr. Hodge became one of the foremost of American geologists, noted for accuracy of observation, truthfulness of statement, clearness of style and unwearied activity. He was probably better acquainted with the Georgia gold field than any one else. He was well versed in Lake Superior geology. He made numerous professional surveys in the coal regions; and the last large work of his life was a topographical and geological map of the consolidated lands of the Cumberland Coal Basin in Maryland. For several years he was co-editor of *Appleton's Encyclopœdia*, and wrote most of the articles on geological subjects, amounting in all to an entire volume of the series. His articles on coal and iron are models of such literature. His right arm became paralyzed, but he continued to work with his left. He spent several years superintending mines in Nevada territory, and in 1872 met his death in a great storm which sunk the steamer in which he had embarked for a port on the north shore of Lake Superior, mourned by a large circle of friends to whom he had become an object of equal respect and affection.

Dr. Henderson, after serving many years on sea and on land, until he was senior surgeon of the United States Navy, was appointed Officer in charge of the Laboratory of New York, the business of which he reorganized and carried on with signal ability, up to the time of his sudden death from an abcess in the ear, in the spring of 1875, aged 58. He had exchanged botany and ornithology, the passions of his youth, for mechanical physics, and spent all his leisure hours, for many years, in discussing the problem of a hot air engine, and in constructing models, in whole or in part, in his little workshop, at the different Naval Asylums over which he had charge when on shore. He was a man of most loveable nature, pure heart and exalted intellect, and was universally respected and lamented.

Mr. Trego became secretary and editor of the Franklin Institute, and afterwards secretary, treasurer and librarian of the American Philosophical Society. His passion for botany and horticulture was practical, and he retained his farm in Bucks county, even during his residence in Philadelphia, where he died at the advanced age of 80 years in 1874.

The sixth year of the First Geological Survey of Pennsylvania was in 1841.

The field work came to a close this year; the expenses of the Survey falling to $12,675.

An Annual Report of only 28 pages was presented February 9, 1842. It stated that field work commenced in May, 1841, with a reduced force of assistants. Mr. Hodge, Dr. Henderson and Mr. Trego had resigned.

Mr. M'Kinley resumed work in the Broad Top coal basin and then upon his map of the Wyoming Valley or Third Anthracite Coal Field, tracing all its numerous anticlinals and studying the collieries.

Mr. Holl completed his work north of the Ohio and west of the Allegheny rivers, and was then ordered to examine special points of geological interest along the entire line of the Juniata river.

Dr. Jackson, with Mr. Ward, made a close and extensive study of the Westmoreland County Coal Field, traced the beds, and established the exact range and force of the anticlinals which part into three basins, south of the Kiskiminetas.

Mr. Boyé was engaged in the Third and Fourth Bituminous Coal regions, south of the Mahoning and north of the Kiskiminetas rivers.

Mr. Lesley, with one companion, made a topographical and geological reconnoisance of the northern counties, from Towanda on the North Branch, keeping in the Fourth and Fifth coal basins, as far as Kittanning on the Allegheny river, and Punxatawney on the Mahoning. He then took up the study of Whelpley's map of the anthracite coal fields, to add to it what had been omitted, or what had been exposed to observation by later mining operations.

Dr. R. E. Rogers continued to make the required chemical analyses of ores, coals, cements, furnace-fluxes, &c.

It will be noticed, that while certain rude limitations of the districts above named kept the gentlemen of the Survey entirely asunder, their respective operations overlapped each other, and overlaid the previous surveys made by themselves or other members of the corps. While it would be hard for any one not personally acquainted with the history of the Survey to assign anything like due credit to the skill and enthusiasm of each, by pointing to any discoveries or generalizations published by the Chief Geologist, yet, the benefit to the State of this method of completing the Survey was indubitably great; for the facts observed by one were sure to be verified or modified in form or statement by the report of another, and generalizations arrived at in one district came into harmony with those arrived at in another. Pennsylvania geology owes its consistency and completeness to the frequency with which every part of the State was gone over, not only by the same, but by different members of the corps. What one eye missed another caught; what one mind mistook the next observer rectified. Not a shadow of doubt or suspicion of grave mistake was left attaching, at the end of 1841, to any important statement to be made by Mr. Rogers in his Final Report; and as a consequence, the Assistants of 1874 remark with astonishment and pleasure how perfectly trustworthy the map and report of 1858 are, so far as they pretend to represent what twenty-six years ago could be observed. The report of the First Survey of Pennsylvania, in this respect, stands quite by itself among American publications of its class and date.

What Mr. Rogers in his Annual Report of 1841 says of the Sixth Bituminous Coal Basin north of the Ohio river, states the most important result of this year's work, and represents the largest generalization arrived at by the Survey. He could have said of all six basins, throughout the length and breadth of Northern and Western Pennsylvania, that "the detailed examinations made during the past year have now made it certain, that the limestones and many of the coal seams are continuous over very extensive areas, though the individual beds frequently change much their thickness and external aspect. The included sandstones, slates and other more mechanically formed rocks display, however incessant variations of thickness and composition, which make it impossible to trace these individually over any very wide extent of country. Thus two seams of coal, in some places separated by sixty feet of strata, will at other points lie within twenty feet of each other." Professor J. J. Stevenson's memoir on the alleged parallelism of coal beds, written at my request, and published with this report, will exhibit the extreme limit to which Mr. Rogers' statement might have reached, had the explorations of Messrs. Holl, M'Kinley and Jackson, in 1840 and 1841, been extended over south-eastern Ohio and West Virginia.

The life-long study of the anthracite basins made by Mr. Sheafer, has revealed the applicability of the statement, in its strongest terms, to that region. The study of the Illinois coals by Mr. Worthen, of the Iowa coals by Mr Hall and Mr. White, and of the Missouri coals by Mr. Brodhead and Mr. Norwood, extend the law of infinite variability to all areas of the Carboniferous Formation. Dr. Le Conte has shown that the law holds good of the Cretaceous (or Tertiary) coal system of the Rocky mountains from Denver City to Santa Fé. It has long been known by Coal Geologists of England. It will be well illustrated by Mr. Platt's report of his Clearfield and Jefferson district; and I will be able to exhibit one of the first steps of its discovery, by reproducing the sheet of Jackson's sections in Westmoreland, which I collated in 1843.

The only other most important result of the field-work of 1841 deserving special mention, was the careful location upon the map of Pennsylvania of the anticlinal axes separating the third, fourth, fifth and sixth Bituminous Coal basins. This was accomplished by the separate action of the Assistants distributed over the whole Coal Area for a distance of two hundred miles. But much uncertainty remained respecting the exact geographical location of the crown of the arch along these anticlinals in the great pine-forest wilderness of Lycoming, Potter, Elk, Clearfield and Jefferson counties; and subsequent surveys have very seriously modified the position and shape of the third, fourth and fifth coal basins in M'Kean county, and in certain local districts of the other counties mentioned above. The general run of the basins is, indeed, exactly that laid down on the Geological State map; but very important practical changes will be made in the blue lines marking the outcrops of the limestones, and in the edges of the dark color marking the outcrops of the Lower Coal Measures, and the distribution of the outlying patches of workable coal beds in Elk, M'Kean and Potter counties.

A third fruit of the sixth season's field-work was the completion of Mr. M'Kinley's map of the Wyoming Coal Basin, upon which he succeeded in portraying almost all, if not absolutely all, of those subordinate diagonal anticlinal and synclinal axes, the number and regularity of which distinguish this anthracite field. The patient toil and carefulness of this geologist, disciplined to his task by the previous mapping of complicated structure on a larger scale—that of the Buffalo and Seven Mountains, west of the Susquehanna river—may be estimated by those who find still useful the Wilkesbarre portion of the Anthracite Map published by Mr. Rogers in the Atlas of the Final Report of 1858. But it must be remembered that that map was engraven in Scotland, and without being subjected to the revision of Mr. M'Kinley, Dr. Whelpley, and its other authors. These gentlemen must not be held responsible for any errors of copyists and engravers which it may exhibit.

The same is true of the Lehigh, Mahanoy, Shamokin and Schuylkill portions of this map.

1842 would have been the seventh year of the Survey, had it proceeded. But the winter of 1841-'2 was passed by the Legislature without the usual annual appropriation for the Geological Survey; and Mr. Rogers was left without sufficient means to put his assistants again in the field, and with all the accumulated wealth of the Survey on his hands, unable to extend it, revise it, arrange it, illustrate it, or publish it.

Piles of annual reports of assistants now called for the authors to become editors; but these had all disappeared, leaving their unrevised manuscripts, their rough sketches, their unfinished maps, their uncatalogued, unstudied mineral and fossil collections behind. No matter which one of these reports the Chief Geologist might take in hand, he was sure to encounter on its first dozen pages a dozen nuts to crack for which he had no cracker. He had acquired, of course, a more or less vague, more or less distinct knowledge of all parts of the State. Only his assistants knew the several parts of the State positively. Thousands of local and personal names stood written in the reports of which he had never heard a whisper, and of which he could make no use for comprehending, verifying or correcting the geographical and geological descriptions. The mere mass of this crude, heterogeneous personal manuscript was appalling. To edit it properly, intelligently, was to repeat in the office the work of the whole corps in the field. It was to task one man with the work of a number of men multiplied by the number of years they had been employed.

Were it not for the lesson to be derived from the inquiry it would be useless now to inquire why Mr. Rogers found himself in this condition,—why the assistant geologists dispersed,—why the Legislature stopped the appropriation. Much might be told and yet leave room for conjecture. "The financial embarrassments of the Commonwealth" was the stereotyped explanation offered by Mr. Bigham's Committee, in their report to the Legislature of 1851*:—

"At the time of the organization of the Survey, it was estimated that it would occupy at least ten years, but the financial embarrassments of the Commonwealth made it expedient and indeed compulsory, to withhold further appropriations after the sixth year, and bring it abruptly to a close before it could be thoroughly completed in all its parts. Anxious to make the

---

*Report of the joint committee of the Senate and House of Representatives of Pennsylvania on the Publication of the Geological Surveys. (13½ pages.)

work as full and symmetrical as possible, the State Geologist continued the exploration, and devoted himself to the preparation of the general final report for three years longer, laboring for the chief part of this period without salary, and at his own expense."

Great indeed must be the "financial embarrassments" of a Commonwealth when its economies descend so low as to decline such inexpensive appropriations for securing the continuance of a national enterprise, and for reaping from it a national benefit.

The explanation is inadequate. In truth, the worth of a Survey of the geological structure and mineral character of the soil of the State to all its citizens was not appreciated, nor comprehended, thirty years ago.

The language of science was then an unknown tongue, and sounded in the ears of the people like the chattering of animals or idiots. The disputes of geologists respecting doubtful points, if listened to at all, were regarded as good evidence of the worthlessness of all their theories; and the truths in which they agreed seemed to clergy and laity alike the insanities of an exalted imagination, or the impious utterances of an irreligious temper.

The mines of the State were (with some most honorable exceptions) bossed by the commonest miners from foreign and quite different geological regions, who had suddenly exchanged the character and position of hewers of coal and pumpers of water at home, for the character and position of mining engineers in America. Ignorant, undisciplined, obstinate, narrow minded and superstitious by nature and habit, and rendered presumptuous and dogmatic by their strange advancement, they were as unwilling to accept as they were unable to acquire a correct knowledge of our geology, so different from their own, and hated professional geologists because these had never lived in childhood, pick in hand, under ground,—because they taught new things hard to comprehend,—and because they denied the propriety of mining the coal of Schuylkill county on the plan of the collieries of South Wales, or of employing the ancient methods of the Cornish tinworks to the brown hematite banks of the Lebanon Valley. The jealousy of professional and "theoretical" interference with traditional and "practical" usages, which has not yet quite disappeared from our mining regions, was in 1842 in all its vigor; and was shared by the landed proprietors, the directors of companies and the general superintendents of collieries and mines. A wave of suspicion and dislike, pushed before it by the First Geological Survey through its whole progress, brought it at last to a dead stop.

The hastily written annual reports of the State Geologist to the Legislature, small in size, fragmentary in order, and necessarily inadequate to exhibit the work accomplished,—indeed not intended for that purpose, but merely to serve as respectful and earnest pleas for an annual appropriation—strengthened the suspicion and dislike of the people of the State towards the survey, and established in all minds, except those few who were in a condition to judge of what that survey was doing, an unconquerable and irresistible conviction that it was a sham and an extravagance—even at the miserable rate of $10,000 or $15,000 per annum.

The Committee Report asserts that "at the time of the organization of the survey it was estimated that it would occupy at least ten years." Some vague suggestion of that sort might have been uttered, but the people of the State never heard of it, and were impatient to have all the counties surveyed at once, and finished up in the very first year. The State Geologist was embarassed by demands of this kind from all parts of the State during the entire progress of the survey, and was compelled to spend his time every winter in the unprofitable task of excusing the great work to the members of the Legislature for not proceeding with an unnatural and impossible celerity.

In spite of the teachings of half a century of scientific exploration people still think that a geological survey can be carried on with flippant and discursive speed. They will wait patiently upon the slow and orderly progress of any other great enterprise, but imagine this to be exceptional. It was no more reasonable that the survey should occupy ten than twenty years.

In its very nature a geological survey is continuous *ad libitum*, and should be perpetual. Its first stages are rapid and of the nature of a reconnoisance, or general survey of the country to be afterwards better surveyed. As it advances it discovers its own future work and prepares itself to do it. The longer it lasts the more local, special, exact and important it becomes. New difficulties arise and fresh problems call for investigation. The publication of what is discovered or determined ought to follow step by step its discovery or determination. To carry on field-work straight ahead, never stopping to revise, collate, compare, calculate, map and publish, is as absurd for a survey as it is improvident for the Commonwealth. But to do this, is to make the survey practically perpetual. If the field-work of the six summer months be followed by six months of office work in winter (and no less will answer) and the results be published immediately for the use of the citizens, there will be plenty of demands for field-work on the same ground the next summer.

It is easy to sketch out the geology of a district, but when the sketch is made the hard work is just begun; the picture is still to be painted. The Reports of Progress in Clearfield and Jefferson counties in 1874, and in Cambria and Somerset counties in 1875, for example, valuable as they will be considered by the people of those counties, are merely complete catalogues of the coal mines, &c., with descriptions of the openings and an index to their geographical position and ownership. The Survey is now prepared to study the relationships of the beds to one another; the interval rocks; their variations in thickness and quality; their extent; their relation to tide water *in detail;* and to show this by a series of accurate local topographical maps and carefully measured sections. And before this slow, close, instrumental part of the work can be finished there will be a fresh development of facts, by new mines and explorations, and by the extension of old mines, which will require still more of this slow, close, instrumental work, and its publication.

In other and more difficult districts, all that has been said above holds true in even a greater degree. There are very many years of work to be done in the whole central belt of the State before a proper and reliable presentation of the facts can be made in all its geographical breadth and in all their geological significance to the public; and it will require quite as many years for the best organized and best trained corps of geologists to work out of darkness into light the extraordinarily difficult geology of the entire south-east section of the State.

Whatever were the causes to bring it about, this was the result: The First Survey was at an end.

For three years Mr. Rogers endeavored to deal singly with an immense mass of material, the accumulated production of several years of labor of many geologists, more than one of whom was his equal in theoretical geology and his superior in field-work, in the vain attempt to reduce this mass to symmetry and illustrate it for publication. That he failed was a matter of course; and no fault of his. He must have been one of the immortal hero-gods of the classic mythology to have succeeded in the effort. To say that he "continued the exploration" of the State was an exaggerated statement of the committee; because one man could do nothing in so vast a field, a field so vast that twenty men had been lost in it for years. And being deprived of salary he was obliged to pursue his profession as a geological expert wherever his clients might direct, but with all the advantages which his official title of State Geologist conferred. The time thus spent and thus paid enlarged his own views and confirmed his reputation at home and abroad, but could go but a little way towards "continuing the exploration" of the State.

Neither could he advance much in the "preparation of the general final report." In fact this was done in Boston in the winter of 1847–'8. There the manuscript reports of the assistants were hastily adjusted together, slightly altered in phraseology here and there, and a copy of the whole made for the Legislature. A second revision for the Edinburgh press was made ten years later, and in this the more important step was taken of altering the nomenclature; substituting *Auroral*, *Matinal*, *Levant*, &c. for formation No. I, No. II, No. III, No. IV, &c. And the further modification of the phraseology was effected; as well as the introduction of new views and new facts obtained by Mr. Rogers during the previous dozen years. To the old report thus modified were added Mr. Lesquereux's report on the botany of the coal; and some general geological chapters on coal, iron, &c. from Mr. Rogers' own vigorous and elegant pen, which stand as monuments of his genius and models for the imitation of geologists who have survived him.

When the corps of 1841 was disbanded, Mr. Sheafer settled as a mining engineer at Pottsville, Mr. Hodge exercised the same profession in various parts of the United States, Mr. Trego and Mr. Ward lived in Philadelphia, Dr. Henderson went into the United States Navy, Dr. Jackson into the practice of medicine at Blairsville, Mr. Holl returned to Europe, Mr. M'Kinley commenced the study of law in Philadelphia, Mr. Boyé practised chemistry, and afterwards became a farmer, and Mr. Lesley indulged in the luxury of a course of theology at Princeton.

Here, at Princeton, chiefly, but also at Philadelphia during the vacations of 1842 and 1843, were put together, on requisition of the State Geologist, the materials furnished by the Assistant Geologists of the Survey for the construction of the topographical and geological, colored, Map of the State, and for the construction also of thirteen long sections across the State, most of which were in 1857 engraved on the same plate with the map.

It is not necessary to describe the difficulties encountered in the construction of this Map. They were due to more than one cause. The materials were of every variety of excellence, from finished manuscripts carefully colored, to the rudest pen sketches on which the limits and characters of the formations and the shapes and directions of the hills and valleys were slightly indicated. In several counties of the State no attempt at all had been made to represent the geographical outcrops of the formations; nor could the attempt have been successful for want of good maps of those counties previous to the work of the survey. The best county maps extant at that time were very imperfect, and as for the State map published by Melish, it was a wilderness of blunders more or less absurd. Not a single county line was in its proper place. The actual latitude and longitude of not one county town had been determined. The largest rivers were erroneously portrayed; the smaller streams were universally fanciful; mountains were made to cross streams, or were omitted altogether; and those which carried outcropping rocks of different geological ages were represented recklessly as continuous or connecting ridges. As a topographical or geographical statement all the northern and western half of Melish's Map was simply a monstrous misnomer.

Yet this was the only available basis for the proposed Geological State Map, and the author had to make the best of the situation. Nor has the Legislature of Pennsylvania taken steps since 1842 towards giving the Commonwealth a good State Map by organizing a trigonometrical, or even an astronomical and telegraphic survey, by which at least the place of some one point in each county could be accurately fixed to serve as a datum pivot of adjustment and rectification for county surveys.

The author of the Geological Map of Pennsylvania, in 1842, had in his hands such of the county maps as had been geologically colored; other county maps, corrected as to the water courses here and there; the manuscript district geological maps of Whelpley, M'Kinley, Henderson and Jackson; and sections of the Melish State Map colored geologically, in whole or in part, and corrected for streams and mountains in certain places, by other Assistants.

To harmonize these heterogenous materials it was necessary, first of all, to discover the amount of county-line and river *errata* in the Melish State Map. There being no reliable lines except those of railway and canal surveys (most of which were inaccessible), and no astronomically or trigonometrically determined points of departure, all the available county maps were brought together, and their errors of county line adjustment* driven from the eastern and western borders of the State towards the centre, and so accumulated along the Susquehanna river. The total error on the lines of parallel being thus approximately known, this error was distributed back again east and west over the whole State; so that the fundamental skeleton of the map was "tempered" like a piano-forte; being erroneous throughout, but with all the local errors reduced to a minimum.

Upon this county-line scheme was then laid down the topography of each county separately, good, bad or indifferent as

*Not one north and south line of a county could be found which fitted the north and south line of the adjoining county.

might happen; corrected in its details by reference to the manuscript geological maps before mentioned, so far as they extended, and by the verbal or pencil notes of the Assistants in districts not geologically mapped. The mountains, which had been the chief guides of the geologists, being carried over from county to county, as the best guides to the corrected representations of the minor water courses.

The mountain belt was then colored in geologically on the basis of the mountain arrangement; the south-east division of the State, from the geologically colored county maps of some of the assistants; and the northern and western bituminous coal area, by reference to the outcrops of the Ferriferous Limestone of the Lower Coal Measures, and the Pittsburg Coal Bed of the Upper, as got from all the reports.

When the map was finished, thirteen lines were drawn across it, radiating from the south-east towards the north, north-west and west; and vertical sections along these lines were constructed to illustrate the contour of the surface and the order and attitude of the rocks at and beneath it. Several of these sections were long enough to cross the entire State; the others, intermediate and shorter, crossed only the mountain belt, and the coal fields.

Several hundred local sections, diagrams and sketches were then drawn, to be used as plate illustrations of the Final Report, or as wood-cuts to be inserted in its text. The majority of these were columnar sections of anthracite and bituminous coal strata, or of single coal beds. Many were mere re-productions of the Devonian and Silurian rock-sections drawn by such of the assistants as had skill and taste for that kind of work. All the columnar sections were drawn to one scale for comparison, and such of the manuscript drawings of rock structure in the reports as needed it were reduced to a common scale.

All this was entirely apart from the production of finished landscape and other drawings by Mr. Lehmann the regular artist of the Survey, and occupied the writer about eighteen months in 1842 and 1843; and it may properly be said that with this the work of the First Geological Survey ended.

In December, 1846, at the request of Prof. Rogers, the writer went to Boston, where Mr. Rogers then resided, and during the ensuing five months made duplicate copies of the State Map, the long sections, and several hundreds of the other drawings, and assisted him in arranging and duplicating the reports. The originals remained in his hands. The duplicates were transmitted to the Secretary of the Commonwealth at Harrisburg. An effort was made to obtain an appropriation at the session of 1848 to publish them, which failed, and they were permitted to lie in the archives of the State.

In 1850, the subject of the publication of the fruits of the First Survey was anxiously discussed by a number of gentlemen, among whom Mr. William Parker Foulke, of Philadelphia, was perhaps the most prominent. With the aid of Mr. Peter W. Sheafer, of Pottsville, these gentlemen succeeded in representing the case so forcibly, that the Legislature was induced to appoint a joint committee of both houses, of which Col. Bigham was chairman, to consider the subject of an appropriation to cover the estimated expense of publication.

An elaborate report was made by the committee and printed in the House Journal, 1851, Vol. II, page 131, and also, separately, in a pamphlet of 16 pages.

A few extracts from this report will serve to explain the next stage of this history, namely: the resumption of field work in 1851.

The whole mass of information and material collected by the Survey, and thus finally systematized, arranged and prepared for the press, was in compliance with the law, deposited in the office of the Secretary of the Commonwealth, early in the year 1847, to await publication by the Legislature. In that position it has been allowed to remain to the present time, its large accumulation of carefully collected and patiently digested facts, bearing directly on many of the most important industrial interests of the State, not yet made practically available to any of the useful ends for which the Survey was undertaken. The knowledge of our mineral possessions, thus successfully secured for the benefit of the citizens of Pennsylvania and the world, has, it appears, cost the Commonwealth an outlay of about $76,657, but by a short sighted economy, this fund of precious information, paid for by the people, is not yet the people's property. It is a valuable capital lying idle a capital which your committee think is capable of yielding rich returns, but it is also susceptible, like much other wealth, of passing to waste by neglect.

Already certain portions of the work, fresh and full of practical utility ten years ago, when the surveys were recent, are behind the wants of the day, and need minute revision.

After giving "a concise and imperfect description of the form and scope" of the proposed Geological Report, the committee go on to say:

In its present shape, it embodies faithfully and fully the results of the Survey as carried on, up to and even beyond the cessation of the appropriations for it. The book and its illustrations actually consist of about three-fourths of the total quantity of material which it is now in contemplation to finally produce.

As we now possess it, the extent of the text is equivalent to about 800 quarto pages. The maps are two in number, but these are both of them elaborate. The quantity of geological sections and other delineations of stratification is estimated at three-fourths of the final amount, while the pictorial illustrations and the fossils are less in the same proportion. The publication at the date of the completion of the Survey, when an appropriation for that purpose was earnestly suggested by the State Geologist, would have been an act of the soundest public policy, promoting in many ways the industrial prosperity of all parts of the Commonwealth. In the interim there has been, in this [the Anthracite] and the other mining districts of the State, a prodigious advance in all our colliery operations in the development of our mineral deposits, contributing to, and at the same time calling for an increase of exactness in geological description.

Your committee recommend that a careful revision shall be made of all those parts of the Survey which the lapse of the time has been making obsolete, and that in its altered and expanded shape, the final report shall be published in a style of accuracy and taste, essential to its scientific character and its practical objects, and creditable to the reputation and resources of Pennsylvania.

It has been ascertained from estimates submitted to a committee of the Legislature in 1848, that an edition of one thousand copies, quarto size, with the maps and other illustrations, in the form of a separate atlas could have been procured in the best style of the art for about $19,000, and an eminent publishing firm, has, at a request from your present committee, presented proposals for executing it in a satisfactory manner, under the indispensable supervision of the State Geologist, its author, and in conformity to conditions suggested by the committee. These proposals stipulate to produce an edition of one thousand copies for the sum of $15,833, in further consideration of the transfer by the State of the copy-right to the publishers. The contemplated enlargement of the book and its illustrations, growing out of the proposed additional surveys and revision, will cost about $3,886, out of the $15,833 of total cost. By this very moderate increase in the expense of publication, your committee firmly believe the true value of the work, as a practical guide to the development of our mineral wealth, and in its usefulness to the whole community, will be more than doubled.

To publish the report in the shape it has at present, without the amendments and revision required by the extensive progress made within the past ten years in developments connected with the coal fields, the iron manufactures, and also the agriculture of the State, would be to give the public a body of statements, in some parts already half superseded by more exact and recent knowledge. Every duty of patriotism to the Commonwealth, and every view of policy to her interests, should prompt the Legislature to publish to the world the fullest and clearest representation procurable, of the resources which she possesses; but to put forth descriptions, some of them which are behind the developments of the day, would be not to proclaim her true resources, but to disparage them.

A systematic re-survey of a large portion of the Anthracite districts will be indispensable, and a revision of the more progressive districts of the Bituminous coal fields west of the mountains, seems scarcely less important.

Among the fruits of such a re-examination will be an increased degree of accuracy in the general geological map, a new geological and topographical map of all the anthracite basins, to supersede the present very creditable one which now accompanies the report, and a minute guide map to the individual coal beds of a large part of the Pottsville basin. There will likewise be produced an extensive series of working sections, such as are felt to be greatly needed by the mining engineer and collier, in their riskful and expensive operations. The distribution and situation of the nearly horizontal layers of coal, iron ore, limestone, and other materials in the western counties, will be yet more fully illustrated than they are now, by an additional set of vertical sections or columns, showing the strata in their several depths below the surface of the county in each neighborhood, and their relative thicknesses and distances apart.

It is easy to see from this catalogue of proposed additions to the mass of geological materials provided by the surveys of 1836 to 1842, that two results would be sure to flow from any favorable action of the Legislature of 1850–'1 upon the report of its committee, viz: 1. The virtual absorption of that mass into a new and larger mass, with the annihilation of all historic individuality of the old work and the old workers; and 2. The setting on foot of a new survey of the State, with a new corps of Assistants, under the former State Geologist, with an indefinite prolongation of field work and an equally indefinite postponement of the date of publication. For no guarantees could be given tó the Commonwealth, as to time, quantity and quality, in 1851, beyond those offered each year from 1836 to 1842. Nor could the Legislators outside of their committee judge of the report even in its then state; much less could they imagine what it would be after further field work. Even the committee were obliged to take the word of the State Geologist for the truth of what they affirmed in their report, which indeed bears internal evidence of having been written by him.

These considerations weighed heavily on the debate when the report came up for acceptance or rejection. Every one lamented that the State Geologist had not originally arranged for the publication of the reports of his Assistants, pure and simple, under their own supervision, annually, during the course of the survey. Every one blamed the Legislature of 1848 for not immediately putting to press the manuscript which the State Geologist transmitted to Harrisburg. But few were disposed to give the State Geologist power to re-inaugurate the survey on its old footing.

But the cry for some report of the old survey, whatever it might be, and under any auspices whatever, which came from all parts of the State was so loud, and the few who knew the value of such a publication were so active in its advocacy, that opposition to the conditions attached to the publication was silenced, and the Report of the Joint Committee was adopted. But the affair had to be carefully managed, and the project for extending the field work had to be kept as much in the background as possible. The paragraphs of the Committee's report, which immediately follow those above quoted, will suffice to indicate this necessity, and they describe beforehand very fairly the actual character of the field work of 1852:—

To represent the geology with this degree of fidelity and close analysis, the re-survey should be conducted with the aid of instruments for measuring and leveling this. More critical research should be facilitated by the labors of a small band of miners, for exposing the outcrops of the coals and other strata. The State Geologist will, therefore, require the services of a corps of geological surveyors. Thus provided, he could complete, it is believed, the revision so much to be desired in two seasons of field examination. It is estimated also, that about two additional years will be then required to enable him to embody the results of these and his other researches into the report, and to supervise the publication of the whole. The already matured portions of the work, such as the topography of the general map, and a large proportion of the more elaborate sections, the engraving of which will involve much time, can be advantageously commenced upon almost immediately. Your committee has ascertained, that upon this plan, the survey can be perfected as far as its practical and scientific objects demand, and the text and illustrations of the report amended and increased, and published with the requisite accuracy in a period of about four years from the present date. The cost to be incurred by the proposed revision, including a salary to the chief geologist for the four years of his supervision of the press and his other services, and the pay to his assistant corps of surveyors, and a small band of miners, for two seasons, will amount, upon a rigidly economical estimate to $16,167. Your committee, therefore, recommend for the revision and publication of the final geological report, a total appropriation of $32,000.

This appropriation was voted and the State Geologist immediately organized his corps of assistants and set them to work in the Anthracite Region, where the largest results were to be expected, and where the most important accumulation of new knowledge had been made, through the mining operations of the ten years which had elapsed between 1841 and 1851. He had himself been employed by the anthracite coal companies as an expert. And Mr. Sheafer, having settled permanently at Pottsville, had become the principal local geologist and mining engineer of the region, and now knew more about its geology than any one else. How the State Geologist himself viewed the situation may be gathered from the next paragraph of the Committee's Report:—

If this thorough revision is authorized, the committee believe that the delay in publication will be fully repaid. A large portion of this time has been devoted by Prof. Rogers in the employ of private companies, to detailed and thorough explorations of portions of the mining districts. The knowledge thus acquired, and the opportunities he has since enjoyed, of comparing his own views on this eminently progressive science, with the opinions of scientific men throughout America and Europe, must largely add to the value of this publication. And, perhaps, it is not out of place for us to add that in addition to all other motives to prompt him to fidelity in the execution of this trust, he is fully aware that upon the merits of this work, his reputation among men of science must stand or fall.

This sentence was intended to be, and was, a complete refutation of all the censures which had been so freely cast upon the State Geologist for the non-appearance of the final report hitherto. He had always shared with his old assistants the two prime elements of a successful scientific character: love of geology as a science, and personal ambition as a geologist In their case the first greatly predominated over the second. With him the two were more equally balanced. Therefore it was folly to charge against him a desire to withhold from the public the results of a splendid train of research, by the publication of which he must needs appear to the scientific world a discoverer and an investigator of the first order of merit. The delay had indeed benefited him in this respect, that it left him to stand alone in the sunshine of this great work; for most of his assistants had dispersed to seek avocations of a different kind, and none of their contributions to Pennsylvanian, that is to say, to Appalachian geology had appeared, or now could ever appear over their own names, or redound to their credit. But for the long delay of such publication the Legislature was alone to blame.

It is not to be supposed, however, that the scientific results of the survey had been buried out of the sight of the learned world because they had not been given by the Legislature to the people of the State most deeply interested in acquiring a knowledge of them. The geologists both of America and of Europe had been kept *en rapport* with the survey, by the visits of Logan and Lyell in 1841, by memoirs read at the meetings of the Geological Association, by private correspondence, and especially by a visit of the State Geologist to Europe in 1848. At the annual meeting of the British Association at Swansea in 1848, Mr. Rogers was the lion of the Geological Section, and there he had an ample opportunity for exhibiting his map and sections, and astonishing his hearers with detailed and illustrated descriptions of geological phenomena grander in extent, and more admirable in simplicity of structure, than anything they had been permitted to study at home.* Nor were his theories less exciting than his facts, delivered as they were with a charming eloquence, seldom equalled and never excelled. The attentions he secured were unbounded and cordial, and confirmed him in his zeal for the publication of the work of the survey in the most complete and elegant shape possible. Two unexpected consequences resulted from this visit after a number of years had passed: his settlement as a professor of science in the University at Glasgow; and the final publication of the work of the survey, in magnificent style, at Edinburgh instead of at Philadelphia.

His acquaintance with the Chiefs of the Geological Surveys of Great Britain and of France enabled him also to draw a just comparison between them and his own, and by no means to the prejudice of that of Pennsylvania. This accounts for an allusion to these surveys in the report of the Joint Committee. After mentioning the expenditure of over $200,000 on the New York State Survey, and before speaking of the resumption of the surveys of Massachusetts and South Carolina, the Committee say:

---

* Report of the XVIIIth meeting of the B. A. F. A. S., held at Swansea in August 1848: London 1849. Transactions of the Sections, page 74, "*On the Geology of Pennsylvania*, by Professor H. D. Rogers. Professor Rogers exhibited a general map of North America, the State Survey of Pennsylvania, and many other maps and sections colored geologically. . . . . Professor Rogers then gave a summary of his theory of the origin of these great parallel foldings in the Appalachian strata. . . . . Drawings were exhibited of the Anthracite Coal mines on the Lehigh river. . . . . . "

The great Geological Survey of France, engrossed the time of two geologists in chief, and a numerous corps of geological engineers of the French School of Mines, for a term of twelve years, in the production of a geological map, and of two volumes of descriptive text.

We may add that Great Britain, sensible of the vital importance to all her industrial interests, of a thorough examination of her strata, has been conducting a close and very elaborate Geological Survey for about twenty years, and that for the past twelve or thirteen years it has been on an organization both extensive and costly, commensurate with the greatness of her mineral wealth, and worthy of her high ambition and eminent practical wisdom.

. . . . . .

A recognition of the utility of this species of investigation, is further shown in the pains taken in several countries to publish periodicals devoted especially to geology and the arts pertaining to it. Such a publication is the "Annales des Mines," issued under the auspices of the French government, and such is the work produced at intervals by the British government, entitled "Memoir of the Geological Survey of Great Britain and Ireland." A further proof of this high estimation of the science, is the fact that the French government supports, at a large expense, her School of Mines, for the especial object of educating and training skillful mining engineers and practical geologists, to aid, whenever called upon, the development of her resources; and so fully awake is the British government to the dependence of much of her industrial prosperity upon her efforts for encouraging practical and scientific geology, that it is fostering with much care a nobly endowed institution in London, called the Museum of Economic Geology, where everything that can illustrate the stratification and mineral wealth of the country, everything that can assist her people to a clear knowledge of the composition of their mineral raw materials, to the best methods of finding, tracing and raising them from the earth, and afterwards converting them to the useful purposes of life, are collected and displayed gratuitously for the inspection of the public, or the special study of the numerous classes needing the information. A library of books on geology and the kindred sciences, on mining and the arts connected with it; a bureau of geological and mining charts, embracing plans and a registry of all underground workings and surveys; also models of the best machinery connected with mining, of the best arrangement for ventilation, and of the various methods in use in different countries for propping, timbering and draining, are here assembled and accessible for study. To these means of information are added a valuable geological museum, of all the rocks and their contents, and the soils derived from them, every kind of coal for fuel, metallic ore for melting, clay for the manufacture of pottery or porcelain, and sands for glass making, and every species of building stone and ornamental rock, in rough and dressed or polished state, is here arranged for inspection; and to add yet further to the instruction they convey, every species of mineral raw material which is transformed in the operations of the arts, is exposed in all its successive stages of manufacture to its highest state of finish.

Two chemical analytical laboratories for the examination of fuels, assay of ores and their indispensable researches, and a noble lecture room as a means of diffusing yet wider to the industrial public, the fruitful facts and principles of geological science and its arts, complete the admirable working machinery of this wisely planned institution. Its respective departments are administered by geologists and chemists, who are among the very ablest in

Great Britain, and it is presided over by one of the foremost of European geologists, Sir Henry de La Beche, who is at the same time at the head of the geological survey of the country, with which this museum of economic geology is intimately connected.

The rest of the Committee's Report (pp. 8 to 16) is an exposition of the practical value of a geological survey in every point of view, including summary descriptions of the coal and iron industries of the State, which need not be here further alluded to.

In 1851 the field work of the First Geological Survey of Pennsylvania re-commenced.

Mr. P. W. Sheafer took charge of the underground work. Mr. John Sheafer headed the transit and level party for cross-barring the surface. M. Edouard Desor volunteered his services for the study of the surface geology. Mr. Leo Lesquereux was the fossil botanist, and Mr. Lesley placed the data thus obtained on paper. Mr. Rogers was himself with the party a considerable part of the season, which lasted until November.

The first part of the season was spent in a thorough transit-survey of the Shamokin and Trevorton coal basins from Ash land westward. The latter part was spent in a similar survey of the Pottsville coal basin from Donaldson and Trevorton on the west to New Philadelphia on the east. The cross-lines extended across the west part of the Mine Hill basin and to the north brow of the Broad mountain.

The work was prosecuted with untiring energy from first to last, and the mapping kept pace with the instrumental work. Long straight transit lines were leveled and staked, from side to side of each coal field, at intervals of 2,000 feet. Longitudinal lines were then run, tying the cross lines-together. With these Mr. Peter Sheafer connected every gangway mouth, shaft, slope and trial pit which then existed, and carried his surveys along these under ground. The terraces which mark the outcrops of the coal beds on the mountain-sides were also run, when needful, and always exhibited on the maps. The mapping was on a scale of about 1,000′ to 1″, so that no feature of the surface was lost. The railways were plotted down, and the principal roads also. Every colliery was designated.

The apparant slowness, however, with which such a work must move on kept the Chief Geologist in a state of impatience, dreading lest the remaining portions of the Anthracite Coal Region could not be finished in the same elaborate style in another year. The corps disbanded at the approach of snow and did not re-assemble in the spring.

In 1852, therefore, the State Geologist engaged a talented Finlander, Mr. Augustus A. Dalson, to do what he could with a single assistant, Mr. Henry W. Poole,* towards a very small representation of the whole Anthracite Coal Field, by reducing the great sheet maps of 1851, and adding to the areas then thoroughly surveyed such an exhibition of the eastern part of the Pottsville basin, and of the Mahanoy and Beaver Meadow basins, as his limited means, and the use of Whelpley's and M'Kinley's maps would permit. This Mr. Dalson did with great skill and taste, and his reduced map appears in the Atlas of the Final Report (1858); but with a wrong title; for to him alone ought the credit of this reduction to be given, the present writer† having had nothing to do with it beyond furnishing the originals from which the closely surveyed parts ot it were reduced. At the same time there could not be a greater mistake committed, than that of saying, as the State Geologist does on page viii of his preface to the Final Report: "The beautiful large map of the Anthracite Region is by Mr. Dalson, *chiefly from his own surveys.*‡"

What was accomplished in other parts of the State during these two years 1851, 1852, was merely nominal, and amounted to little. But the years 1852, 1853 and 1854 were very important for the unwearied efforts of Mr. Rogers to revise his immense wealth of facts, and to leave no new important develop-

* The inventor of the euharmonic organ.

† The latter, indeed, never saw the map until it was published, and was in nowise its author.

‡ In the paragraph in which this statement is made, no mention is made of Mr John Sheafer, who headed the surveying party. Mr. W. B. Rogers, Jr., is also spoken of as the only other assistant *geologist* with Prof. Desor. Whereas he was then a student, and was not seen by those who were engaged on this survey of 1851 at all. Mr. P. W. Sheafer is only quoted as a "surveyor (performing also geological functions)"; whereas he was *the* geologist of the survey of 1851 *par excellence*, knowing more of the field than all the rest of us combined.

ment of geology in the State unnoticed. He visited most of the principal iron mines, and especially investigated the classic region of the fossil ore about Danville. His memoir on this, included in his Final Report, is masterly. But one man can do little or nothing by himself in so great and varied a field of scientific labor, however signal his zeal and ability. The First Survey of Pennsylvania may be truly said to close in 1851. All between that and the appearance of the Final Report in 1858 was of the nature of superficial revision in the field, and editorial labor in the cabinet. There was no museum; no arranged collections; no students of the minerals and fossils of the State. Mr. Rogers was, himself, neither mineralogist nor palæontologist, and never professed to be. His structural geology also had all been provided for him by the old survey. But let any one read the special memoirs with which he closes the second volume of his Final Report, and there can be no sentiment but one of admiration for the breadth of his views, and the clearness, force and elegance of his delineations. No geological paper has ever appeared, excelling, in every good scientific quality, his memoir on coal.

To form a safe estimate of the diligence of the State Geologist during these years it is only necessary to observe that, with the exception of Mr. Lesquereux's beautiful memoir on the Coal Plants (pages 837 to 884), 275 quarto pages, in fine print, at the end of the Second Volume, viz: from page 698 to page 1022, are entirely from his own pen, and represent the devotion of more than twenty years of observation and reflection to some of the grandest problems in geology.

In 1855 the Final Report was expected by the terms of the act of 1851. The State Geologist had asked but four years, and they were out; but the book was not. Why not—the following Report of a Select Committee of Inquiry, appointed by the Senate of Pennsylvania, will sufficiently explain. The document is too valuable not to be placed on record, uncurtailed in this history:

Report of the Select Committee of the Senate of Pennsylvania in relation to the progress and present condition of the State Geological Survey.

The committee appointed to inquire into the progress and present condition of the State Geological survey, and what steps are necessary to complete the publication of the report, submit the following statement:

In the year 1851, the State Geological survey, which owing to the financial embarrassments of the State had been suspended for a number of years, was, in consequence of the earnest and repeated desires of a large portion of the Commonwealth, brought to the consideration of the Legislature, with a view to the final publication of the great and valuable mass of materials that had been collected. The Legislature, after having been placed in possession of the history of the work and the causes of its suspension, in an elaborate report, (to be found in House Journal, 1851, vol. 2, page 131,) made an appropriation of thirty-two thousand dollars, for the revision of such portions of the field work as, from the rapid development of the mining regions of the State, required re-survey, and for the publication of the report itself, with the accompanying maps, plans, cuts and sections, in a style suitable to the requirements of the work, and the reputation of the Commonwealth. A joint committee, consisting of two members of each House, with the then Secretary of the Commonwealth, was authorized to issue proposals for a contract, under which the publication of the work might be made in accordance with the design of the Legislature, (Pamphlet laws 1851, page 636.) Several bids were received, from which the committee selected that of Hogan & Thompson, of the city of Philadelphia, as deemed most favorable to the interests of the Commonwealth, and the proper publication of the work. The proposals upon which the bids were founded, and the contract ultimately entered into, contemplated the application of eighteen thousand dollars to the revision of the field work, including the indispensable services of the State Geologist, and fourteen thousand dollars to the publication of the report. The last instalment of eight thousand to be retained until the completion of the work, which it was anticipated would cover a period of four years. The geologist was to receive his portion of the appropriation from the publishers, who were to draw and disburse all the funds. Under this arrangement the geologist, Professor Rogers, commenced his field work early in the spring of 1851, and continued with a small band of assistants actively engaged, until the close of the season suspended his field labors. The firm of Hogan & Thompson having intermediately been dissolved, and its successor being compelled to suspend payment, no part of the appropriation was paid by them to the State Geologist, according to the agreement. In this embarrassing position, and having consumed his own means in the expectation of immediate reimbursement, he appealed to the Legislature for redress. Through the aid of a special committee, a settlement of his account was effected with the publishers, and an arrangement was made, by which, for the future, he was authorized to draw that portion of the appropriation intended for the field work, directly from the treasury (Pamphlet laws 1852, page 388.) At this period, the spring of 1852, a large amount of material, consisting of the general geological map of the State, sectional drawings, &c., was ready for the press, and was amply sufficient to occupy all the available resources at the command of publishing houses of the highest ability, for fully two years. The geological map was placed in the hands of the publishers for engraving, but they, although urged by the geologist, neglected to put him in communication with the engraver selected for the purpose, although the necessities of the work, and the terms of the agreement between the publishers and the geologist, alike required the constant personal supervision of the latter over the work, as it passed through the engraver's hands. Meanwhile, the publishing firm which had succeeded that with which the committee originally contracted, was finally dissolved by the death of the senior partner, and another firm took its place, which, after a short-lived existence, soon likewise disappeared. No commercial firm then remained even in *name*, to represent the parties with whom the Commonwealth had contracted. Although the publishers had drawn, on account of their contract, the sum of four thousand dollars over and above the amount paid to Mr. Rogers, they do not appear to have made any progress towards the publication of the work.

The time limited for its completion, will expire on the 14th day of April, 1855. Your committee, therefore, addressed letters to the original firm and their successors, inquiring as to the progress of the work and probability of its completion. Answers have been received from most of the parties interested, which admit that nothing has been done, and amount to a virtual abandonment of the contract. From the information before the committee, they are induced to believe that the four thousand dollars so advanced to the publishers, and for which the State has received no equivalent, is totally lost, as no sufficient security was taken to cover these advances.

In this recital, your committee have not felt themselves called upon to make any particular reference to the want of proper management, which has resulted in this loss to the State, and the delay in the publication of so valuable a work. The duties of their inquiry will be best discharged by concluding this brief history of the matter, with some suggestions as to the most effectual mode of extricating the work from its present difficulties, and securing its early and complete publication, according to the original design of the Legislature.

Of the fourteen thousand dollars appropriated by the act of 1851, for the publication of the report, the sum of four thousand dollars was drawn, as already stated, by Hogan & Thompson; and of the eighteen thousand dollars, intended to cover the expenses of revising the field work, the sum of fifteen thousand dollars has been drawn by Mr. Rogers. The latter gentleman has exhibited vouchers for the whole amount, and has been engaged from April to the middle of November of last year, in prosecuting, with his corps of assistants, the field researches. By perseverance in his field and closet labors, the State Geologist has now a large accumulation of prepared matter in readiness for publication. All the geological and topographical maps are finished; so likewise the pictorial illustrations of the work, and nearly all the geological sections, columns of diagrams, sketches and drawings of the fossil organic remains. A schedule of these materials is annexed. The text of the final report, intended to be, in many of its parts, entirely re-written or much expanded, to embrace the results of the minute revision the survey has undergone, is likewise in an advanced stage of preparation, and can readily be completed before it will be practicable to enter on the printing of the letter-press. Treating, as the work does, of features, where scrupulous accuracy and minuteness of delineation is indispensable, it will need the immediate and constant supervision of the author in nearly every portion; and it is believed, upon a careful estimate, that it will require about three years for its faithful execution. The difficulty of having such a work properly supervised in its progress of publication, is abundantly evident in the embarrassments which have impeded it thus far. The policy of imposing such a duty on a joint committee of the Legislature, as heretofore, is very questionable, as it cannot be supposed that they can give it the attention requisite; nor, indeed, that they would be likely to possess the peculiar qualifications for such a task.

In view of these facts, it has occurred to your committee that they can suggest no more prudent or economical plan than to confide the whole work, in its supervision and publication, to the State Geologist himself. With him it has been the work of a life time; and his professional reputation is involved in the manner in which the results of so much labor and research shall be spread before the world.

Mr. Rogers, at the invitation of the committee, has offered to assume the contract upon the same terms as heretofore agreed upon between the State and Messrs. Hogan & Thompson. The proposal of this firm, at the time, was about the same as those of other publication houses of established means and character; but it is believed that the increased value of material, labor and artistic skill, during the last four years, will enhance the cost of such a publication nearly twenty per cent. The chief inducement to Mr. Rogers, in undertaking the task, would be in the subsequent control of the copy-right of a work, in which his professional, as well as his pecuniary interest would be so extensively embarked. He offers the most ample security, and is willing that the State should protect itself in the time and mode of payment. These considerations, together with the character of Mr. Rogers, and his qualifications for such an undertaking, afford an ample guaranty for the faithful performance of the contract. His written proposal, dated February 10th, 1855, is hereto annexed. From this it will appear that no further payment will be required than the balance of the appropriation made by the act of 1851, excepting the sum of four thousand dollars, to replace the amount advanced to Hogan & Thompson.

There is, however, one alteration that your committee would recommend. The great map will be the most important feature of the report; for it is intended to exhibit not only every geological, but every geographical and topographical feature of the State. In order to embrace so vast an amount of detail, it has been found necessary to reduce the work to a fineness that none but the most skillful artists can execute. By enlarging this map to twice its present scale, it will correspond in size with Barnes' map of Pennsylvania, published in 1854, under the supervision of Wm. E. Morris, Esq., and will exhibit clearly and distinctly to the eye, the minutest results of the Survey.* The variations in the physical character of the State can be faithfully portrayed; not merely the mountain ridges, which contain such incalculable mineral wealth, but the benches and spurs of those mountains, with the mining drifts and collieries. The State could not present to her citizens the results of this Survey, in a more useful or acceptable shape; and the increased expense will be no objection, as the State Geologist is willing to undertake to produce a map of double the present scale, for an addional charge of two thousand dollars.

With a view of relieving this matter of its present difficulties, and of accomplishing the object above indicated, your committee submit a supplement

*This sentence embodies two errors: for the map was not enlarged to double the size, but from $7\frac{1}{2}$ to 5 miles to the inch as can be seen below; and secondly most of the "minute details" on Henderson's map were not and could not be exhibited.

to the several acts relating to the geological survey, and authorizing the Secretary of the Commonwealth to contract with Mr. Rogers for the publication of the report, upon the same terms as agreed upon by the former contractors.

N. B. BROWNE,
FR. JORDAN.

HARRISBURG, *February 10, 1855.*

To the Committee in charge of the Geological Survey of Pennsylvania:

GENTLEMEN:—Influenced by a strong desire to see the Geological Survey published, and believing, after much examination of the matter, that this cannot be accomplished so profitably to the true interests of the Commonwealth, as by my assuming the task myself, I hereby avow my willingness to accept the contract for publishing the final report of the said Survey, consisting of the materials defined in schedules formerly presented, on the same terms as were heretofore agreed upon between the State and Messrs. Hogan & Thompson, provided this shall not interfere with the payment of the balance still due me as State Geologist for services under any contract or appropriations.

I wish to be allowed three years for the deliberate and careful execution of this responsible duty, of producing, in the highest style of accuracy, this patiently elaborated and permanent picture of mineral wealth and geological structure of Pennsylvania.

In accepting this contract, I would forego the salary to which I should otherwise be equitably entitled, for indispensable services in supervising and virtually editing the work, if awarded to another publisher; and I will observe that this renunciation of salary makes my offer, compared with all former bids by other parties, cheaper, at the moderate rate of salary I have received, by four thousand five hundred dollars.

In reply to your inquiry respecting the cost of enlarging the general geological and topographical map of the State, to twice the scale upon which it has been prepared, I would say that it will require, upon an approximate estimate about two thousand dollars additional appropriation to re-construct and engrave the map in this larger form. That this suggested expansion would enhance its value for both geographical and geological reference in a much higher proportion than the additional cost would measure, no intelligent person, understanding the wants of the community in the matter of an exact topographical map of the State, can for one moment doubt.

Very respectfully,
Your obedient servant,
HENRY D. ROGERS.

Statement of the materials of the Final Report of the Geological Survey, now ready for engraving, &c.

1. General Geological and Topographical Map of Pennsylvania, (1 inch=7½ miles.)
2. Topographical Map of entire Anthracite region, (½ inch=1 mile.)
3. Topographical and Geological Map of Pottsville coal field, (1⅝ inch=1 mile.)
4. Small Map of coal fields and avenues to market.
5. Small Map of lead and copper district of Chester and Montgomery.

Geological Sections.

1. All the detailed sections of the coal strata of the Pottsville basin illustrating the positions, dips, distances apart and relative thicknesses of the several coal beds for every two or three miles from Mauch Chunk to Dauphin, embracing about twelve principal sections and a number of subordinate ones. (To be reduced in scale in the engraving.)

2. A series of similar detailed sections, illustrating the structure of the great middle coal fields, the Shamokin, Mahanoy and Lehigh basins, about eight principal sections, with several intermediate local ones. (To be reduced in scale.)

3. A similar suite of exact sections of the coal strata of the Wyoming and Lackawanna basin, about eight chief ones with several lesser intermediate sections. (To be reduced.)

4. About ten principal and several smaller *vertical columns* of the strata, showing the thicknesses and the intervals between the coal beds in the several anthracite basins.

5. A series of sections across the bituminous coal fields of the northern and western counties.

6. An extensive suite of vertical and columnar sections of the bituminous coal measures, illustrating the stratification in some fifty important localities.

7. About fifty diagrams of individual coal seams, anthracite and bituminous, showing the sub-divisions, qualities of coal and slaty partings in each bed.

8. Diagrams illustrating the courses of outcrops of coal beds, and of gangways in mines.

Sections not pertaining especially to the Coal Fields.

1. A series of fourteen elaborate general sections, traversing the State from S. E. to N. W., to elucidate the general map.

2. A series of more local sections, explaining the stratification of districts on the Delaware, Lehigh, Schuylkill, Susquehanna, Juniata, Kiskiminetas and Allegheny rivers.

3. A suite of sections showing, in detail, the structure and stratification of the counties S. W. of the Susquehanna, and S. E. of the Allegheny mountain.

4. Several local sections, illustrating the geology of the lower or S. E. counties.

Pictorial Illustrations.

1. Ten water-color views of scenery, to be engraved on copper, or color-printed on stone. (All ready.)

2. A series of outline sketches, thirty in number, of Topographical and Geological features, and of mines, quarries and mine fixtures, to be etched on stone or copper. (The greater part now ready for the engraver.)

3. Several elaborate pictorial sections, or panoramic geological views, including a full one on the Schuylkill from Philadelphia to near Norristown. (Ready for engraving.)

Drawings of Fossils.

The thirty plates of drawings of organic remains mentioned in the schedule of material rendered in 1851, are now finished, and awaiting the best artistic skill of a scientific lithographer.

Text.

Of the text, about five hundred pages of descriptive matter, pertaining chiefly to districts which have undergone a close revision, are written. Now that the chief mass of the field work is finished, the uncompleted parts of the revised text, with other remaining closet work, can proceed, while the Geologist is bestowing a portion of his attention to the supervision of the engraving of the extensive body of illustrations designated in the foregoing list.

Respectfully submitted by

H. D. ROGERS.

The recommendations of the Committee were adopted, and the contract transferred to the State Geologist. All the documents of the survey became his personal property; as well as the copyright of the book and atlas when published. One thousand copies were distributed among the members of the Legislature, and found their way into public and private libraries, and to second-hand book stores. Each of the Assistants received a copy. The rest of the edition belonged to the State Geologist, who arranged for its sale by Johnston in Edinburgh, and Lippencott & Co. in Philadelphia. Several hundred copies remained unsold at his death, and what remain of these may be purchased of Van Nostrand in New York.

The two quarto volumes of the Final Report contain 1631 pages of clearly printed matter, with 778 beautifully engraved wood-cuts inserted in the text; 7 engraved plates of geological sections, uncolored; and 23 plates of coal plant fossils, engraved by W. & A. K. Johnston, Edinburg. These are bound at the end of the second volume.

There are also 6 double page engravings of sections along the rivers of the State, and 2 three-page plates of vertical sections, bound in or at the end of the first volume.

Beautiful colored lithographs of three of Lehman's finest water-color paintings are also given. One of the most striking—the Pulpit Rocks at Huntingdon—forms the vignette to Vol. I; another equally fine—the old Baltimore Company's outcrop near Wilkesbarre—the vignette to Vol. II; and the third—a very effective interior, looking towards the entrance—faces page 382.

These, however, are the only reproductions in colors of a long series of exquisite water-color drawings, made by the talented artist, for the survey, most of them easel pictures, fit for the walls of any museum in the world. The scenery of no region has ever been more magnificently represented, and that with as much faithfulness to its natural history, as skill and taste in color and perspective. These paintings would have made a large and charming art gallery for the Capitol at Harrisburg. Whether they are scattered now, never to be re-collected, or whether they remain in the possession of Mr. Rogers' heirs, is not known.

Besides these numerous water-color paintings, Mr. Lehman, during the years of his engagement on the survey, executed a still more extensive suite of pencil and pen and ink sketches of scenery in all parts of the State, illustrative of its topography and geology. Twenty-three of these are reproduced as one page plates in Vol. I, and twenty in Vol. II. This represents only a small selection from the whole portfolio, and of such as were best fitted by their shape to appear on a quarto page.

Besides all this, there are two maps in the second volume; one at page 675, of the mining district of Chester and Montgomery counties; and one at page 1019 of the anthracite and bituminous coal fields of the State.

A description of the wealth of facts presented and illustrated in this remarkable book—one of unexampled excellence in its day, reflecting the highest honor upon authors, editor and artist—would be in vain attempted here. But the plan of its arrangement should be stated, and the book itself left to the research of its readers.

Description of the Final Report of 1858.

Volume One commences with an introductory memoir on the Physical Geography of Pennsylvania, pages 1–57, describing the boundaries, water sheds, anticlinal, synclinal and monoclinal mountains, their scenery and structure, their crests and gaps, the river basins, rain fall, temperature and prevailing winds—of the different districts of the State. Lehman's sketches of the gneiss rock mass at the mouth of the Wissahiccon,—of the Kittatinny (Silurian) mountain (and two of its gaps) as seen from the Summit mine,—of the east end of the Bald Eagle (Silurian) mountain, as seen from the west,—of the high synclinal end of the Terrace (Devonian) mountain, as seen from Huntingdon,—of the Tamaqua anthracite coal basin, as seen from the Summit mine (in the foreground),—of the Pass of the Lehigh through the Second (Devonian) mountain below Mauch Chunk,—of the Upper Coal Measure country about Canonsburg in Washington county,—and of the bend of the Susquehanna around the dying synclinal (Silurian) Bald Eagle mountain at Muncy—illustrate this Chapter in a perfectly satisfactory manner. These illustrations are only equalled by recently published sketches of the Rocky Mountain surveys.

The Geology of Pennsylvania is then introduced (p. 59–60) by a short classification of the rocks into gneissic or hypozoic, and palæozoic.

PART I. *Metamorphic Strata* of Pennsylvania. An introductory chapter (62–66) classifies these along the Atlantic Slope of the Middle and Southern States; and then the gneissic rocks of Pennsylvania proper, into three districts.

Chapter 1. The *Southern Zone* of Gneiss, south of the Chester Valley (pages 66–82). This is sub-divided into three belts, a southern, middle and northern, crossing the Schuylkill river above Philadelphia, and the Brandywine west of West Chester. A running landscape section of the left bank of the Schuylkill* shows the outcrops from Spring Mill down to Manayunk bridge, with the mills and other buildings against a background of hills.† Another detailed section is given along the Wissahickon creek.

---

* Drawn by Mr. Dalson?

† This section must not be taken without much caution as a basis for restoring the structure, for the gaps are long and numerous, and the variations of *strike* are not shown. The exposures along the opposite right bank are even more important, and the two series must be combined. No map of the strikes was made; and this is an indispensable prerequisite for so difficult an investigation.

The text, also, descriptive of this interesting section, is open to serious criticism; and the classification of the gneiss into three belts needs revision. The error seems to have been committed of regarding the gentle north dip at Philadelphia as a nearly horizontal *cleavage plane*, instead of *plane of stratification* (p. 70), which supposition renders any attempt at an interpretation of the structure fruitless.

Another great defect in the section, as given, is the absence of *south* dips, for a long distance north of the soapstone quarry. These make the structure between the quarry and Spring Mill a high sharp *anticlinal, dying rapidly eastward*. This was not comprehended, because no map was made; and because the theoretical view, entertained at that time, took for granted *a continuous ascent in the order of formations* from Philadelphia to Spring Mill.

Chapter 2. The *Middle Zone* of Gneiss, north of the Chester Valley, and south of the New Red country (pages 83–90.) The difference between these (now known to be Laurentian) gneisses and the soft (Huronian ?) gneisses of chapter I is well set forth, and the iron ores of the district are alluded to.

Chapter 3. The *Northern Zone* of Gneiss, of the South Mountains (pages 91–103.) The minerals of Chestnut Hill north of Easton, the Highlands south of the Lehigh as far as Reading, with a section along the Delaware river (98), and unconformable contacts of Primal on gneiss (100), and other sections across the range towards and at Reading (102, 103) are described.

PART II. The *Palæozoic*, or ancient fossiliferous strata of Pennsylvania, occupies the remaining pages (104–586) of Vol. I, and the first 666 pages of Vol. II.

INTRODUCTORY BOOK.

Chapter 1 (104–109) gives a short synopsis of the twelve great formations, from "Primal" to "Seral," including the Coal Measures,—with their New York Survey synonyms, and their characteristic qualities and thicknesses in various critical places in and out of Pennsylvania.

Chapter 2 (110–120) sub-divides the Palæozoic region into nine districts, and the districts into many belts.

Chapter 3 (122–146) gives a summary description, or bird's eye view, of the twelve great formations and their sub-divisions; their lithological characters; and tables of varying thickness in different parts of the State.

BOOK I. *Primal and Auroral Strata of the Atlantic Slope; (pages* 149–236.)

DIVISION I. *Primal Series.* Introduction.

Chapter 1. *Primal Series*, southern belt. This describes the supposed Potsdam sandstone and slate No. 1, along the south

---

And yet the resemblance of the northern and southern rocks is noticed; and some astonishment is expressed (page 73) that the "northern zone," half a mile wide on the Schuylkill, should "run to a point before reaching the Valley of the Wissahickon."

A hasty preliminary map survey of the dips and strikes along the river, made by Messrs. Young and Fagen in 1874, sufficed to show that this chapter must be re-written, and to prove that in the early days of the world's history a double crested mountain range extended from Trenton, past Philadelphia towards Baltimore, the height of which, on the present cross-line of the Schuylkill, equalled that of the Swiss Alps.

side of the Chester valley; first, generally, east and west of the Schuylkill river, and then in detail; with the four anticlinals east of Willow Grove; the quarries; the belts of serpentine south of the valley, and the serpentine and chrome iron ore on the Maryland line (149–171.)

This chapter is illustrated by 12 wood-cuts, mostly sections, some of which show the plicated structure of the district very well; but it is evident that a great deal of hard work has to be done in Philadelphia, southern Montgomery, Delaware, southern Chester and southern Lancaster counties, before the geology of this corner of the State can be really understood. Nothing could well be more meagre and unsatisfactory than our knowledge of it, embodied in this chapter of the Final Report of 1858. Nor can it be comprehended until a close *instrumental* survey of every part of it be made. Its limestones, serpentines and iron ores must else remain in the mystery in which this chapter leaves them. It is evident that the *theory of structure* here given is, as a whole, utterly unreliable and ir many important respects wrong.

Plates 3 and 4, at the end of the volume, show the exposures along Brandywine creek.

Wood-cut, fig. 20, is a good illustration of *cleavage;* wood-cut, fig. 15, as an illustration of *surface creep* acquires now a new interest in the light of the discoveries (1874) of glacial action around Philadelphia, to which action the bent outcrops may be ascribed.

Chapter 2. *Primal Series*, bounding the middle gneissic district around Waynesburg in northern Chester county; in the North Valley hill and Mine Ridge; in the Welsh mountains; and in Chiques rock, at Columbia; (173–182.)

Wood-cuts 21, 22 and 23 are beautiful folded sections across the Chester valley and North Valley hill. Wood-cut 24 gives a very doubtful theory of the Chestnut Hill ore region at Columbia, and 24*a* an admirable illustration of the way our large hematite ore beds are sub-divided, massed, and crushed underneath loads of contorted tinted clays. A large amount of instrumental work is needful to cast light on the difficult geology of the large district thus summarily disposed of in this short chapter.

Chapter 3. *Primal Strata* on the Susquehanna and in York county, (184–195.)

Plate 5, line 1, at the end of the volume, gives the exposures along the left bank or Lancaster county side, from the Chiques Furnace to Columbia.

Plate 6, of seven long lines, gives all the observed rock-exposures along the right bank or York county side, from the Maryland line up to Wrightsville, opposite Columbia. These sections are in the main excellent, and show the minute care bestowed upon the south-east problems by Prof. Rogers after the close of the survey in 1842. But their *true* interpretation waits for a complete *instrumental* survey of the York and Lancaster border country, by which alone the numerous inland outcrops can be properly referred to the river banks, and the actual structure be learned. In 1874, Professor Frazer commenced this work, and his section along the west bank assists in the interpretation of a part of the line. At least two years of hard work in York and Lancaster is needed.

The geological description of York county in this chapter is extraordinarily meagre, and amounts to little more than the re-production of the scanty notes published in one of the annual reports.

Chapter 4. *Primal Rocks* of the South mountain, between the Delaware and Schuylkill, (196–202.) The sandstones at Easton, Bethlehem, Allentown and Womelsdorf are located, and the outcrops indicated geographically.

Chapter 5. *South Mountains*, south-west of the Susquehanna, (203–207.) Into these five pages are thrown a few cursory notes, all that the First Survey had to say of this large and most important district of the State. One wood-cut (fig. 25) illustrates its supposed structure along the line of the "Gettysburg railroad," or pike from Gettysburg to Chambersburg. The total ignorance of Huronian and Laurentian geology which prevailed even as late as twenty years ago made the problem of our South Mountains insoluble. But the real explanation of the failure of the First Survey to elucidate this geology is to be sought in the fact that no *instrumental survey* was ever made of these mountains. This is one of the tasks commenced by the Second Survey.

DIVISION II. *Auroral Series.*

Chapter 1. General description of the Chester Valley limestones; trap dykes; marble quarries; iron ore banks; and the insulated limestone basins of Lancaster county with their iron ores, (209–219.)

Here is inserted the double page plate of the exposures along the Schuylkill river; and at the end of the volume Plates I and II repeat this section in a less picturesque manner, but in greater detail and on a true scale. Four pictures are given (on one page, opposite 213,) of quarry faces of limestone and slate, illustrating admirably the *cleavage* in all its type forms. These are very instructive. The same thing is shown in sketches accompanying Fig. 27, page 218.

Chapter 2. *Auroral limestone* in the isolated basins of Lancaster, York, Adams and Chester counties, with trap dykes, iron ores, &c. (220–231) and little wood-cuts, (figs. 29 and 30) of Nevin's quarry west of the Brandywine, and the broken saddle of gneiss pushing up the limestone at Bailey's between New London and Kennett Square. In this chapter we have a good deal that we miss in Chapter 1 of Division I.

Chapter 3. *Auroral Limestone* of the South mountains, between the Delaware and Schuylkill, with the zinc mines of the Saucon, (233–236.) The similar synclinal limestone valley south of Carlisle is not alluded to here.

BOOK II. SECOND DISTRICT. *Kittatinny Valley.*

Chapter 1. Boundaries; dimensions; cleavage; typical formations (F. I and F. II)—Auroral Magnesian limestone; Matinal limestone; Matinal black slate; Matinal newer slate; fossils; structure; (237–240.)

Chapter 2. *Auroral Limestone* from Easton to Reading. Section (text) along the Delaware. Cleavage. (241–245.)

Chapter 3. *Matinal Slate* from the Delaware to the Schuylkill, (246–249.) Here is inserted wood-cut 31, showing overturned and compressed folds in the slate near Harrisburg, very instructive. Wood-cut 32 exhibits the roofing-slate cleavage at the Delaware Water Gap. The quarries at Slatington are described; and the surface covering of boulders remarked, but assigned to diluvial instead of to glacial action.—Hydraulic lime.

Chapter 4. *Auroral Limestone* from Reading to Harrisburg. All that is told of it is on page 251.

Chapter 5. *Matinal Slate* from the Schuylkill to the Susquehanna, with its iron ores, tongues of limestone, shore-breccia at Bunkers' Hill, and outlying Hole mountain, (252–256.)

Chapter 6. *Auroral Limestone* from Harrisburg to Greencastle, (257–259.)

Chapter 7. *Matinal Slate* from the Susquehanna to the Maryland line, page 260.

Chapter 8. *Auroral Limestone* and *Matinal Slate* belts in Franklin county, (261, 262.)

Nothing could be more unsatisfactory than the description as above given of this most important belt of the State. But what else could be expected from a survey forty years ago, not organized on an instrumental basis, unable to dig or bore, and at a time preceding the opening of most of the iron ore mines of the Great Valley. The few mines which then afforded opportunities for observation are sketched in the report.

Chapter 9. *Iron Ores* of the Kittatinny valley, (263–267;) in which not a section or diagram is given; and almost the only important contribution to our knowledge is that of the borings at the Copperas Mine, near Trexlertown.

To make up for the deficiency of this chapter, Prof. Rogers has discussed the iron ores of the Great Valley, with those of the country to the south of it, in the light cast upon them between 1840 and 1855, in a separate memoir, extending from page 712 to page 740 of Volume II. But even this is of little value. The Second Survey has therefore taken up the investigation *de novo*, and as if nothing at all had been done in the Great Valley, and hopes, by making an elaborate contour line map of the entire belt across the State, to arrive at some clear statement of its geology, especially in relation to the iron ore deposits.

BOOK III. *Stroudsburg-Orwigsburg Valley.*

Chapter 1. *Limits of the district and character of the formations*, embracing the Silurian and Devonian, up to the Coal, *i. e.* from the Oneida, No. IV, to the Catskill, IX and X; in other words, all that belt of valley land lying back of the Blue, Kittatinny, or First Mountain, and in front of the Catskill, Pocono, Mauch

Chunk, or Second Mountain, from the Delaware river and New York State line to the Susquehanna river above Harrisburg, (270–279.)

Chapters 2 and 3 (280–281) give short descriptions of the belt along the Delaware, with a section at Walpack Bend, (fig. 33.)

Chapter 4 (283–287) continues it to the Wind Gap, with charming cuts of structure at Stroudsburg, (figs. 34, 35, 36.)

Chapter 5 (287–289) continues it to the Schuylkill with a curious diagrammatic comparison of the curves at the Wind Gap and Lehigh Water Gap (fig. 37) and a totally false section of Stone Ridge at the Lehigh, (fig. 38.)

Chapter 6 (289–292) continues it to the Swatara, with a section along the Little Schuylkill (fig. 39) showing the series of anticlinals, but not showing the fault in the Blue mountain, (fig. 39.)

Chapter 7 (293) continues it to the Susquehanna, and fig. 41 shows the difference in contour produced by a difference in dip.

Beautiful sections through the Delaware and Lehigh Water Gaps are given on lines 5, 6, 7 of Plate V at the end of the volume. In 1875, Mr. Chance has made more perfect studies of these gaps, with contour line maps of them, and they will be modelled in plaster and colored, as type specimens of this feature of our geology. He has effected important structural changes in the section north of the Lehigh Gap; and Mr. C. E. Hall has added glacial phenomena; but the sections on Plate V are nevertheless excellent. On the same plate, line 3, the Gap of the Susquehanna is shown; but a much more elaborate section is given in fig. 60, on page 328. And, opposite page 365 is Lehman's fine landscape sketch of the ends of the mountain, separated by the broad river.

BOOK IV. *North-Eastern District*, (294–312); occupied exclusively by Devonian rocks.

Chapter 1 (294–302) sketches broadly the Vergent, Ponent and Vespertine formations, their thickness, lithology, ripple marks, false bedding and fossils.

Chapter 2 (303, 304) gives the structure and a few local details of the Delaware river (Catskill) highlands.

Chapter 3 (305, 306) gives the structure and some local details of Wayne and Susquehanna counties.

Chapter 4 (307, 312) sketches the Bradford and Tioga county finger-mountains, the Wellsboro' and other valleys, the Mansfield limestone and iron-ore-belt outcrops (page 309, and fig. 43); a vertical section on the Tioga river (fig. 42); the remarkable wide spread (glacial?) deposit of New York limestone-boulders over the country (page 310); R. C. Taylor's (Vespertine) limestone-outcrop south of Wellsboro' (fig. 44); and a remarkable lake and (glacial moraine?) ridge on Crooked Creek (312.)

These chapters are extremely meagre; and therefore Mr. Sherwood was appointed to make, in 1875, a systematic survey of this Fourth District of the First Survey. He will be prepared to hand in, this winter, a geologically colored map of Bradford and Tioga counties, illustrating his report, in which the more important (Chemung) horizons are carefullv traced and described.

BOOK V. *Lower Juniata District.*

DIVISION I. *Perry and N. W. Franklin counties;* the first and second synclinals, back of the Blue or North mountain, and in front of the Tuscarora mountain, (313–366.) We here enter on the carefully worked up mountain belt of the State; and in Dr. A. A. Henderson's old district.

Chapter 1. Sherman's Valley. North Horse Valley. Burn's Valley. Path Valley. Amberson's Valley. M'Connellsburg Cove.—The structure is carefully described; with the great fault along the north side of Path Valley (319,) and its iron ores (322); the great fault opening the Cove; Lowry's Knob; and the fossil ore shales of Little Scrub, Dickey's, Cove and Little Cove mountains.

Henderson's Figs. 45 to 51, of the Path Valley fault and iron ore, are curious and valuable, although subsequent mining operations, and private surveys, have revealed features hidden in 1839, tending to modify somewhat his conclusions. Fig. 52 repeats the fault at the Carrick bank with a different interpretation.

Henderson's Figs. 53 to 59 tell the whole story of the M'-Connellsburg fault beautifully. The shattered fragments of Oneida (No. 4, Levant white S. S.) caught in the fault, as shown in figs. 54, 55, 56, are specially noteworthy.

Chapter 2. *Levant, Surgent and Scalent* (Oneida, Medina, IV, Clinton, &c., V,) formations in Perry and Franklin counties (328–345), in detail. The *Levant* is shown in a section of the Blue mountain at Harrisburg (fig. 60), carefully drawn, but without the mathematical elements, and therefore only general; and the Surgent and Scalent series are sketched on page 329.

Levant Series; geographical distribution (330–336.) In this Henderson gives his mountain system, as it occupies his map:—Tuscarora mountain, Conecocheague (North) mountain, Bower's mountain, Round Top, Dividing mountain, Rising mountain, Amberson's knob, Clark's knob, Jordan's knob and Parnell's knob. The Kittatinny, North or Blue mountain is traced and described minutely (334–336) from Harrisburg west to Mercersburg, with its hooks, gaps and anticlinals.

Surgent Series; geographical distribution (336–345.) In this Henderson shows all the outcrops of the Clinton, (No. V,) red shales holding the fossil iron ore, of which, however, he has very little to say, because it was at that early day scarcely any where opened. This is almost the only deficiency in this part of the Final Report of 1858; and this deficiency is being remedied by the Surveys of 1874, '5 and '6, by the minute detailed examination of these outcrops by Mr. Dewees. Henderson mapped the undeveloped outcrop of the formation with great care in Pfout's, Raccoon, Shaeffer's, Kennedy's, South Horse and Horse valleys, and along the flanks of all the Levant (No. IV) mountains, above mentioned.

The structure of the region is given by Henderson's cross section, in figs. 61 to 63, on page 345.

Chapter 3. *Scalent, Premeridian* (VI) *and Meredian* (VII) strata in Perry county, (345–354.)

The marls and limestones of VI, (Niagara, Lower Helderberg, &c.,) are described by Henderson together, and traced by him through all their meandering convolutions, in Pfout's valley and Raccoon valley, (346) and along and across nineteen anticlinals numbered from south to north, (346–350.)

The Oriskany sandstone ridge, (No. VII, Meridian,) which carries these limestones on its flank, and is bent backward and forward by these anticlinals, is described with the same care,

(350–354); the anticlinals and synclinals being beautifully drawn in wood cuts 64 to 68.

Chapter 4. *Cadent and Vergent* (VIII) strata in Perry county, (355–362) in Wildcat ridge, Raccoon ridge, Half Fall mountain, Dick's ridge, Mahanoy ridge, and Fishing Creek valley; with the iron ores of Juniata Furnace (358) and Perry Furnace, (360) and at Oak Grove, (362.) The anticlinals and synclinals of the district are again exhibited in section in Henderson's cuts, figs. 69 to 71, page 362.

Chapter 5. *Ponent* (Red Catskill, IX) rocks in Perry county (363, 364), with the anticlinals and synclinals again shown in figs. 72, 73. Here come in notices of Berry's mountain, Buffalo mountain, (Middle ridge, Hominy ridge, VIII,) Cove mountain and Peters' mountain.

Chapter 6. *Vespertine Conglomerate* (White Catskill X) and *Umbral Red Shale* (XI) in Buffalo mountain, in Hunter's valley and in the Cove, north of Harrisburg; with a notice of the north end of the wonderful, thin, trap dyke which crosses the Susquehanna river at Duncannon, (the mouth of the Juniata,) and may be traced in a straight line southward to Boiling Springs south of Carlisle, cutting all the formations. More of Henderson's cross-sections, with anticlinals and synclinals, are given in figs. 74, 75, page 366. They have no special reference to this chapter; lying west of the district, and showing the fault in Path valley. But no other arrangement of the wood-cuts was feasible, inasmuch as they cross all the formations, while the text follows each formation along its outcrop. By this arrangement of the report reference to the sections is made difficult and tedious. But it was unavoidable; or Dr. Henderson would have devised a better method.

DIVISION II. *Tuscarora Valley.*—This is the third great synclinal belt, and includes the third great anticlinal belt of Shade and Black Log mountains, extending from the Susquehanna, in Union county, to the northern part of Fulton county (367–395.) The fossil ores of this range will furnish the staple of Mr. Dewees' Report of Progress for 1874–'5.

Chapter 1. *Levant* (IV) in East Shade mountain and Blue Ridge.

Chapter 2. Black Log Valley (II and III); Western Shade

mountain; Meadow, Cook's, Shade and Pott's gaps. The extraordinary cross fault, filled with iron ore, mined at the crest of the Black Log mountain, south of Orbisonia, is alluded to on page 371.

Chapter 3. The Tuscarora valley in detail; its Clinton red shales and fossil ore of Slenderdale ridge, the Juniata Long Narrows; Lost Creek valley; the fossil ores at Mifflintown; on Forge ridge; in Licking Creek valley; Kurtz' valley; Tuscarora valley proper (378) and its sub-dividing ridges and fossil ore outcrops; Pfout's valley (380); Thompsontown; Waterloo (381); Burnt Cabins; the M'Connellsburg cove fault (again; 381.)

Chapter 4. The limestones (VI) of Tuscarora valley, along the Slenderdale ridge and over the same ground again to the M'Connellsburg cove, (383–387.)

Chapter 5. The Oriskany (VII) ridges, (388–390)

Chapter 6. The Slate hills of VIII, from the Susquehanna river across the Juniata, through Tuscarora, Kurtz's, Pfout's and Little Aughwick valleys; with their Devonian iron ores (390–395.)

No wood-cuts accompany any of the chapters in this division; because Henderson's sections cross his entire district, and are given in the previous division of the report.

DIVISION III. *Lewistown Valley.* This remarkable belt of low land between mountain walls extends from Selinsgrove on the Susquehanna river, past Lewistown, into Maryland, and formed the northern limit of Henderson's field, and the northern border of his manuscript colored map. It was the first ground occupied by the Mr Dewees' party in 1874, and will be richly illustrated by contour-line maps of Mr. Billin, and the instrumental cross-sections of Mr. Ashburner. Mr. Lehmann's landscape sketch of the Blue hill on the Susquehanna at Northumberland, and Dr. Henderson's cross-section at Lewistown (fig. 77, page 406), his little cross-section of Dry valley (fig. 79, page 409), and his two minute map-cuts of the limestone knobs at Beavertown, and of the topography of Forge ridge, are the only illustrations of its geology given in the Final Report.

Chapter 1. Character of the formations:—Surgent; fossil ore; Scalent; marls; limestone; Premeridian; Meridian (VII) shale; sand; Cadent; older black slate; ore; Vergent; Ponent, (IX) (396–401.)

Chapter 2. Topographical features:—Meridian (VII) ridges through Dry valley; at Lewistown, &c. (401–403.)

Chapter 3. Flexures or Axes:—Jack's; Shade; Moser's valley; Adamsburg valley; at Lewistown four axes described (407), with their steep ridges, iron ore, limestone, &c. (403–409.)

Chapter 4. Iron Ores:—Only one page is given to the fossil ore of V, and the hematites of VIII, (410.)

Chapter 5. From Lewistown to Maryland; in the Lewistown (Juniata) valley; in Pigeon cove; in Little Scrub ridge; ending with a slight notice of the fossil ore in the Little Cove.—Sulphate of barytes is mentioned (414) in the red shale of V, (411–416.)

Chapter 6. Limestones of VI throughout the belt, (416–420.)

Chapter 7. Sandstone of VII throughout the belt, (420–424.) The Oriskany (Meridian) ore at Chester furnace is noticed, (421.)

Chapter 8. Shales and Sands of VIII throughout the belt (424–429); ore of the Green Briar valley, (426.)

Chapter 9. Catskill (IX and X) at the south-west end of the belt in the great Aughwick valley and Big Scrub ridge, (429–431.)

Chapter 10. Iron ores of the district (431–433.) Only two pages are given to this subject, and only a description of the small operations at the old Matilda, Hope, Hanover, Chester and Brookland furnaces.

It may be said that the Fifth Book of the Final Report of 1858 is a very exact reproduction of Henderson's statement, very full as to the complicated structure and lithology of the region, but almost destitute of information as to its lime, sand and iron ore treasures, and therefore of the very highest scientific value, but of very inferior practical importance. It is from a practical or metallurgic point of view that the present survey regards its work in this district.

BOOK VI. *Upper Juniata District*, (434–575.) From Jack's mountain to the Allegheny mountain, and from Danville to

Cumberland. This great interior belt of Silurian and Devonian formations, two hundred miles long by fifty wide, studied by M'Kinley and Jackson, is divided into two anticlinal and two synclinal zones.

DIVISION I. *Montour Ridge Anticlinal Zone.*

Chapter 1. Levant, Scalent and Surgent of Montour ridge, with its fossil ore;—Premeridian, Meridian, Cadent and Vergent of Union county and the Muncy hills, (434–440.)—Descriptive section at Danville (435.)—Vertical section at Muncy (439.)

Chapter 2. Montour ridge in detail, with three cross-sections (figs. 81, 82, 83) showing the iron ore beds, and vertical section, at Danville (fig. 84.) The Danville mines (443.) Montour's ridge further east and west. A thorough discussion of the quantities of hard and soft ore in the ridge.—This is one of the most carefully written and valuable chapters in the report (441–450.)

Chapter 3. The Devonian flanks of Montour's ridge (451, 452.)

DIVISION II. Fishing Creek anticlinal valley (Devonian,) north of the Third Anthracite coal field.

Chapter 1. Upper Fishing Creek (453, 454,)

Chapter 2. Muncy Hills synclinal (454–456.)

Chapter 3. Buffalo valley, west of the Susquehanna (456–460.)

DIVISION III. Third Belt, between Jack's mountain and Bald Eagle mountain; the Buffalo, and Seven mountains, Stone, Tussey, &c.

Chapter 1. General structure (461–469.) Lehmann's beautiful outline pictures of the terrace-bordered Kishacoquillas valley, and a view of Black Log and Shade mountains from Mifflintown are here given.

Chapter 2. Character of the formations (469–475.) The Auroral and Matinal of the valleys, and the Levant of the mountains, with a descriptive section at Bellefonte, are carefully defined.

Chapter 3. Kishacoquillas valley, in detail, with its folds, and coves, and keel-shaped partitions, and terraces, its rocks, and its iron mines, (475–480.)—Lehmann's striking view of the mountains of IV at the east end.

Chapter 4. The Seven mountains, (480–483.) Three of M'Kinley's cross-sections (figs. 85, 86, 87) show the anticlinals.

Chapter 5. Buffalo, Nittany and Bald Eagle mountains (483–489.) Short, Brush, Nittany, Big mountains are defined; Sugar, Pheasant, Oval and Nippenose valleys are explained, and two cross-sections of them are given (figs. 91, 92.) Longitudinal views of the way these looped mountains are drained through ranges of gaps (figs. 88, 89, 90) are very instructive. The ravines in the south terrace of Nippenose valley are finely shown in fig. 94. Lesley's sketch of the "eddy hill" in Anti's gap of Nippenose valley (fig. 93) is re-enforced by Lehmann's beautiful page landscape of the Lehigh Water Gap, with a similar (glacial?) mound in the foreground. This picture ought to have been given with the text of Book II, chapter 3.

Chapter 6. Nippenose valley (490.)

Chapter 7. Sugar valley (491), with M'Kinley's cross-section.

Chapter 8. Brush valley (492), with a section of its unsymmetrical axis (fig. 96.)

Chapter 9. Penn's valley (493.)

Chapter 10. Nittany valley (494–498.)

Chapter 11. Iron Ores of Nittany valley (499); naming the old banks. This is a very poor description; but without a good local map little more could be done. A complete map of this ore region, and its continuation to Bedford, will be published by the present survey.

Chapter 12. Nittany valley continued westward (500–503.) Lehmann's picture of the Natural Bridge in Canoe valley, illustrates the erosion of all these limestone valleys.

Chapter 13. Canoe valley (504–506) with a side view of the Tussey mountain terrace (fig. 97.) Lehmann's fine landscape view of the Indian Chief Rock near Williamsburg, shows tne cleavage.

Chapter 14. Morrison's cove (507) with Jackson's section of the Bloomfield ore outcrop (fig. 98) is good for nothing. It does not show the fault, and is otherwise imperfect.

Chapter 15. Snake Spring valley and Friend's cove (508.)

Chapter 16. Bean's cove (509) fig. 99.

Chapter 17. Will's mountain and Millikin's cove (510) brings the belt to the Maryland line, and is pictured grandly by Lehmann. On the same large plate is placed Lehmann's charming sketch of the famous view from the hill-top back of Lewistown, showing the entrance to the Long Narrows, the terraces of Black Log, and the great anticlinal mass of East Shade mountain. With these pictures the division appropriately ends.

DIVISION IV. *Broad Top Synclinal.*

Chapter 1. Stone valley (512–521); its anticlinals running into the Seven mountains and the curious spurs of Tussey mountain; the fossil ore (V) of the mountains; the Oriskany sandstone (VII) of Warrior's ridge; its pulpit-rocks, illustrated by diagrams (figs. 100, 101); and pictured in Lehmann's two landscape sketches of Stone ridge on the Juniata below Huntingdon (facing pages 519 and 520), and in the beautiful colored vignette plate to this volume, by Lehmann.

Chapter 2. Hare valley (521); with a diagram (fig. 102) of the duplex form of Jack's mountain.

Chapter 3. Woodcock valley (522–527); with three cross-sections (figs. 103, 104, 105) The more recent developments of the fossil ore by R. H. Powell and others are not described.

Chapter 4. Tub-mill harbour (527.)

Chapter 5. Broad Top (528–533); with one small cross-section of no value, fig, 106;—Trough Creek valley and its ore of XI;—descriptive section of the Red Shale mass of XI. This slight sketch amounts virtually to ignoring this important coal field; and little is added to it in Volume II devoted to coal. In fact no geological description of Broad Top has ever appeared, although private surveys have made it very clear, and large private maps will be published by the present State Survey in 1876.

DIVISION V. *Fifth belt along the Allegheny mountain.*

Chapter 1. Upper Silurian formations at Milton, Muncy, Williamsport and Jersey Shore (534–541), with a vertical section of the same at Cumberland on the Potomac for comparison, (fig. 108.)

Chapter 2. Upper Silurian, between Williamsport and Bellefonte (545–552); with descriptive sections and cross-sections (figs.

109, 111); a curious diagram section through the Bald Eagle mountain to show its *belts of vegetation* (fig. 110); and Lehmann's landscape view of Bellefonte and the gap.

Chapter 3. Devonian between Muncy and Bellefonte (548–552); with an instructive cross-section of a group of anticlinals on Larry's creek (fig. 112); and another through Short mountain, showing the place of the Vergent ore lifted by the Canoe run anticlinal (fig. 113.)

Chapter 4. Silurian and Devonian from Bellefonte to Hollidaysburg (552–555); with the fossil-ore outcrops at Hannah furnace in section (fig. 114); a cross-section and excellent close description of the limestone strata at Bell's Mills (fig. 117.) A set of lengthwise drawings by Lesley of the effects of changes of structure on the topography of the Bald Eagle mountain is given in fig. 115; and his map diagram of the theory of the formation of the Hollidaysburg hatchet, in fig. 116. Lehmann's magnificent picture of the terraced end of the Canoe mountain, as seen from Sinking valley, faces page 555.

Chapter 5. Silurian and fossil ore south of Frankstown and Hollidaysburg (555–558); with a curious drawing of the displacement of Canoe mountain at the fault (fig. 118); and a landscape sketch from Blue Knob, of the anticlinal end of Dunning's mountain at Hollidaysburg (fig. 119.)

Chapter 6. Lower Helderberg, Oriskany and Devonian, south from Hollidaysburg (558, 559); with three cross-sections (figs. 120–122); and Lehmann's fine landscape view of the Allegheny escarpment, with its high terrace knobs and ravines, taken from the rocks above Hollidaysburg.

Chapter 7. The fossil ore of V in Dutch Corner near Bedford (560); with a cross-section. A good report of this important ore region will be given in 1876.

Chapter 8. Devonian in Blair county, along Chestnut ridge, (561, 562) with a section and description of Lime ridge near Bedford (fig. 124.)

Chapter 9. Devonian south-west of Buckstown (563–565); with two more of Jackson's cross-sections: at Buckstown (fig. 125); and on the Bedford pike (fig. 126); with a diagram section (fig. 127) to show the normal form of the escarpment of the Allegheny mountain.

Chapter 10. Bedford Synclinal Basin to the Maryland line (565–570); with Jackson's rude sketch map of the Bedford neighborhood (fig. 126) and his cross-sections (figs. 129–134), showing the plications under the town and the larger anticlinals of the region.

Chapter 11. Country west of Buffalo mountain (570–575); with Jackson's sections (figs. 135, 136) showing the larger features of the structure, but utterly worthless for any close local study of the geology in detail.

The whole of the district thus described, or rather sketched, in this book of the report, deserves a thorough instrumental survey.

BOOK VII. *Seventh district in the north-west corner of the State.*

Chapter 1. Erie, Crawford and Warren counties, where the Devonian formations rise from beneath the Bituminous Coal field and border on Lake Erie (576–581.)

Chapter 2. Economical Details of the same (582–586.)

The few pages given to this region show plainly how barren a field it was regarded previous to the discoveries of oil and gas; previous also to the creation of a demand for the bituminous coal of Western Pennsylvania by the western manufacturers of New York, and by the mining and metallurgical industries of the Upper Lakes.

BOOK VIII. ANTHRACITE COAL REGION (Vol. II, p. 1–465.)

Introduction (1–21.) Position of the anthracite basins in the mountain chain; the surrounding Vespertine (X) mountains, and Umbral (XI) red shale valleys; stratification of the Coal measures; Seral Conglomerate (XII), Lower Coal and Lower Barren measures, Upper Coal and Upper Barren measures; with a typical cross-section, fig. 137, (16) and a detailed description of the great conglomerate..

DIVISION I. *Southern, or Schuylkill Region* (26–235.)

Chapter 1. General structure and sub-divisions (26–41.) Lehigh and Little Schuylkill district, Schuylkill and Swatara district, Mine Hill basin, Wiconisco basin and Dauphin basin;—the basins A, Aa, B, C, Bc, Cb, Cc, c, cc, and anticlinals *A*, *AB*, *C*, *b*, *c*, *cc*, between Mauch Chunk and Patterson; basins CD, D,

Dd, DF, DE, EF, F, FH and GH, FG, f and ff, ff and fff, FFF and FFFF, and their corresponding separating anticlinals between Patterson and Pottsville;—basins H and I, Ii, Ji, JK, K, and axes, between Tuscarora and Mill Creek;—basins C and E west, EF, FG, GH, Hh, hI, iJ, JK, between Mill Creek and West Norwegian;—the western series of flexures; anticlinals L, M, *L* (or F west), LM, N, MN, O, P, *PP*, *QQ*, Q, *Q*, R, *R*, RR, RRR, S or K, QR, RS, *S*, between Pottsville and Tremont.

This will give some idea of Mr. Rogers' system of notation, by which he endeavored to overcome the extraordinary difficulties attendant upon a detailed geographical and geological description in language of the plications which, on the map, are indicated by long and short straight black lines. It requires patient attention to his system of lettering, both in the text and on the map, to enable the reader to comprehend the local geology.

All the other anthracite coal fields are provided with corresponding sets of letters to distinguish their plications; but local geographical names, better known to miners and residents, are also freely used in the report. It is perhaps a pity that these local names were not exclusively adopted, and mere numbers used on the map to refer to them.

Chapter 2. Nesquehoning and Panther Creek (or Tamaqua) district (47–75.) Mauch Chunk end of the basin; synclinal belts; Nesquehoning mines, with cross-section (fig. 138); coals in tunnels 1 and 2; coal seams; Rhume Run mine (figs. 139, 140, 141); Summit mine, sub-divisions shown in Vertical section (fig. 142);—Local types of the coal measures east of Tamaqua;—groups and principal coal beds; shutes at Mauch Chunk; tunnels. A little picture of a cave-working in the great bed at the Summit mine forms a tail piece to Chapter 4, Division II, Book VIII. And Lehmann's landscape view on page — shows the Summit mine to great advantage. An ideal cross-section of the basin is given on line 1 of Plate VI, at the end of the volume. Mr. Rothwell's recent work supersedes this.

Chapter 3. Tamaqua Coal Measures (76–83), with vertical sections of beds C, D, E; and cross-section of Greenwood tunnel.

A noble cross-section at Tamaqua is given on the first line of Plate I at the end of the volume.

Chapter 4. Schuylkill District (84–96.) The anticlinals are again described, to show what beds are lifted to the surface, and especially the Peach mountain axes.

Chapter 5. Tamaqua to Silver Creek in detail (96–119). Twelve cross-sections exhibit the order and attitude of the coals in the basin, and seven vertical sections show the individual coal beds. Lehmann's picture of the Newark Tunnel colliery gives a fair idea of the country and the mode of reaching the coal, above and below ground. A long cross-section from Middleport to the summit of Sharp mountain is given on line 2 of Plate I, at the end of the volume.

Chapter 6. Silver Creek to Mill Creek in detail (119–140). Whelpley's beautiful section of the roll in Potts and Whites' tunnel (fig. 167 on page 127) is a fine expression of the plicated and squeezed condition of many of the coal beds in the central belt of the basin. Two other sections cross the basin. A beautiful full cross-section of the basin along Silver creek is given on line 3, Plate I. A still longer and finer cross-section from Port Carbon up Mill creek, through the Mine Hill gap to the summit of Broad Mountain, occupies lines 4 and 5 of Plate I. All these are on the basis of Whelpley's work, filled in with details subsequently obtained.

Chapter 7. Mill Creek to West Norwegian in detail (141–152). The collieries are taken in order and discussed, with two cross-sections. Here is given, in tabular form, the description of the outcrops in the Pottsville gap, a fine vertical section of which is produced in Plate II at the end of the volume. Seventy-four strata are described. Whelpley's three long sections from Pottsville across the basin up East, Middle and West Norwegian creeks, and a section of the Gate Vein measures in Pottsville, occupy lines 1, 2, 3 of Plate II.

Chapter 8. West Norwegian to Swatara in detail (152–167), describes the anticlinals, troughs and coal beds on West and West West Branches. Fig. 173 gives the Otto tunnel. Whelpley's sections along the West Branch and Wolf creek, at Minersville, and through Silvertown, occupy the 4, 5, 6 lines of

Plate II. Other smaller sections on Swatara creek waters occupy lines 1 and 2 of Plate III.

Chapter 9. Collieries between West Norwegian and Swatara (168–179) on Wolf and Middle creek. Fig. 174 gives a curious sketch of the twist by which the Bast and Pearson slope leaves one bed and takes another in its descent. Fig. 175, Serrill's tunnel coals.

Chapter 10. Middle creek to Tremont and Donaldson (179–188). Two tunnel sections, and two vertical columns illustrate this in the text, and a long section at Donaldson appears in Plate III, with some smaller ones. Whelpley's beautiful cross-section of the west end of the Pottsville basin, showing the overturned Devonian and Carboniferous in Second and Sharp mountains on the left, the four sub-basins and three anticlinals at Tremont and Donaldson, and a landscape back-ground, or bird's-eye-view, westward, of the ranges of mountains, and the forking of the Pottsville basin into the Dauphin basin on the south and the Lykens Valley (Wiconisco) basin on the north, is one of the most striking illustrations in this volume (fig. 176, page 180.)

Chapter 11. Wiconisco and Dauphin basins (188–198) general description; then in detail; with four superb synclinal cross-sections (figs. 181–184). Cross-sections of Red mountain at Tremont; Sharp mountain coals at Rausch and Swatara gaps; a long section on Rausch creek from Little Lick mountain to Sharp mountain, with the Red mountain axis in it; and fine cross-sections at Bear gap and Klinger's gap, occupy lines 3, 4, 5 of Plate III at the end of the volume. These, and other sections above mentioned, were perfected or wholly constructed by the work of 1851.

Chapter 12. Mine Hill basin (198–212) and first a description of the Mine Hill and its anticlinal, which splits this basin off from the Pottsville basin proper. Four cross and one vertical section come in the text (figs. 184–188.)

Chapter 13. Important coal beds and their local names (212–225). This is a most valuable index to the collieries; and the local and duplicate names of some of the best beds are given on all the ground above described. Of course the list is susceptible of great enlargement, now that twenty years of mining operations have elapsed since this chapter was written. It is equally true that the descriptive text of all the foregoing dozen chapters of this book must be hereafter modified in its minuter details, and a few important errors be corrected. But the principal part of the work is done and its description stands firm. What is now required is to bring up our knowledge of the underground geology to date. A thousand valuable facts have been revealed by the lengthened gangways and tunnels, and by the deepening of slopes and shafts. No essential modification, however, is possible. All that the sinking of the great shaft of the Reading Company has proven, is predicted in the descriptive text and cross-sections of these chapters, although it was an important event, the *proving* of Whelpley's assertions in 1839; and a still more important event, the getting by this shaft a true section of the measures in the deep, for no surface study possible could define the *exact* shape of the tongued flexures in the centres of the sub-synclinals, nor the exact amount of distortion of the crowns of the sub-anticlinals, underground. The laudation of M'Ginnis for predicting the underlay of the Big Bed where he sunk his trial shaft, was infinitely absurd, for the whole thing was already published in detail in the report of 1858, if the Pottsville people had taken the trouble to read it. The excitement at that time, and the delighted astonishment of the coal-world at the more recent success of the Reading Co.'s shaft, are our best evidences of the sad fact that Mr. Rogers' Final Report of the abundant accurate work of the First Geological Survey in the Anthracite Region has, in reality, never got into the hands of the owners, operators and miners of the Schuylkill region,—*has never been seen or read.* It is a great pity. But the book was too bulky and too costly to get into many hands, and if some idea of its contents will induce the owners and operators of the anthracite coal region to clean out Van Nostrand's stock of the remaining copies of the edition, so that one or two hundred copies may be distributed among the mining engineers, superintendents and boss miners of the region, or be placed where they can be studied, the object of this sketch of one of the most remarkable books the world has ever seen will be reached.

Chapter 14. Small coal basins on Broad mountain (225, 226).

The two very pretty but worthless cross-sections, figs. 189, 190, furnish all the information got up to 1859. These basins were considered very unimportant. It is now known that they deserve a thorough survey and elaborate description.

Chapter 15. Classification of Pottsville coal beds (226–233.) This chapter is Mr. Rogers' own production, and he alone must be held responsible for its accuracy. It is illustrated with the following *combined vertical sections of the coal measures*, viz:

Fig. 191. Coal strata at the Little Schuylkill.

Fig. 192. Coal strata on Silver creek.

Fig. 193. Coal strata on Mill creek.

Fig. 194. The lower part of the same for comparison.

Fig. 195. Coal strata on West Branch and Wolf creek, and

Fig. 196. Coal strata at Donaldson; Thick mountain; of which sections the principal feature is, of course, the Big Bed near the bottom of the measures, 55 feet thick in fig. 192, and sub-divided by many feet or yards of parting rock in the other sections. The old error of making two great beds, the "Mammoth" and the "Jugular Vein" is here shunned. Mr. Rogers neither letters nor numbers his beds, but gives the accepted local names. Mr. Daddow's attempt to transfer from the Bituminous region and apply to the Anthracite coal beds the letters of the alphabet was a mistake and a failure; and it may be useful to remark here that all that is valuable in Daddow's Book on Coal was taken from Rogers' Report of 1858, with the addition of chapters filled with geological nonsense, against the influence of which upon their minds students and miners should be warned. The author is now dead; therefore severer criticism is improper. But a needful lesson is taught by the fact that, had the report of the First Survey been published in a cheap and handy form and become thus universally distributed and discussed, it would never have been supplanted, in its rôle as a text book of geological knowledge for the citizens of Pennsylvania, by a mischievous travesty, disgraceful to the author and the age.

Cha ter 16. Faults and Slips (234–238). This chapter, furnished with facts and drawings by Whelpley, shows the settling of the upturned rocks of the Sharp mountain, sliding on each other (fig. 197); the dying of an anticlinal (fig. 198); a swell and crush in a coal bed (figs, 199, 202, 203), from actual instances; and the modes in which the flexures break and slip, doubling the outcrops (figs. 200, 201.)—Mr. Rogers writes his own beautiful chapter on this subject, expanding it so to take in all the principal phenomena, on pages 885–916 of this volume—(See below.)

DIVISION II. *Beaver Meadow Region.*

Chapter 1. General Description of this Eastern Middle Anthracite Coal Field (239–243.) The anticlinals of Green mountain, Dreck's ridge, Council ridge, &c. &c. and the Beaver Meadow, Hazleton, Black Creek and other basins.

Chapter 2. Beaver Meadow Basin (244–251) with two sections of the Big Bed (figs. 205, 206) and two little cross-sections.

Chapter 3. Hazelton Basin (252–260) with three poor little cross-sections, and two sections of the Big Bed. M'Kinley's long sections from Hazleton south to Spring mountain, and across all the basins through Hazleton, are given on lines 1 and 2, Plate IV, at the end of the volume.

Chapter 4. Black Creek Basin (260–264.) Whelpley's seven original little cross-sections (figs. 215–221) merely hint at the structure. A world of exploration has been done in this and the Hazleton basin since 1851, none of which appears in this report, and it may be said not to really describe the region in any proper sense of the term. A new and thorough survey of the whole plateau is called for.

DIVISION III. *Mahanoy and Shamokin or "Western Middle" Coal Field.*

Chapter 1. *Mahanoy coal region* (265–306.) General structure and topography. The three Mahanoy and the three Shamokin basins, with their subordinate troughs and anticlinals, are described and illustrated by some of Whelpley's neatest little drawings, cross-sections, showing merely the conglomerate and cover of coal measures, such as the end of Head mountain (fig. 225); its two basins (fig. 226); the eastern origin of Bear Ridge axis (fig. 227); and four others (figs. 229–232) showing its increasing importance further west; his general section across the region (fig. 224); and his lovely little broken anticlinal crest (fig. 228.)—There are also sections of the beds cut in Bear Ridge tunnel (fig. 233); through Green Ridge gap (fig. 236); and Shamokin gap; and eight columnar sections of the Mammoth bed and one or two others.—The subsequent instrumental survey of 1851 furnished the four long elaborate sections across the Mahanoy and Shamokin basins, presented on Plate IV, and the two magnificent sections at Trevorton on Plate VI.

Chapter 2. West of Shamokin gap (306–317). Trevorton and the Zerbe's Run basin, surveyed by Mr. Rogers before 1851, illustrated by five cuts of the coals in the text and two noble sections at Eyster's gap, on Plate VI. An interesting woodcut on page 317 shows the slate roof of Coal 10, near Trevorton, covered with a mat of flattened stems of sigillaria.

DIVISION IV. *Wyoming Coal Field.*

Chapter 1 Form and features of the Valley of the Northern Anthracite coal field (318–330); general view of the strata, with six of M'Kinley's vertical sections:—at Cobb's gap; Solomon's gap; Nanticoke; Kingston; Wilkesbarre; Mill Creek (figs. 251–256); general structure of the basin; undulations; Campbell's ledge conglomerate, plicated slightly (fig. 257); Coal Measures; sub-divisions of the field:—1. Carbondale, 2. Scranton, 3. Pittston, 4. Wilkesbarre, &c. and 5. Wyoming, &c. which are taken up in that order and described in detail in the following chapters.—Plate VI, at the end of the volume, gives M'Kinley's long cross-section of the basin opposite Warriorpath gap.

Chapter 2. Carbondale and Archbald district (330–337); Lackawanna valley, upper end; with vertical sections of the Great Bed at Carbondale (fig. 259) and two other beds; and of the coal measures at Archbald (fig. 260.)

Chapter 3. Scranton district (338–360); with its undulations; the Roaring Brook basin; Pine Brook basin; Diamond Coal mines; between Hyde Park and Pittston; between Springbrook and Scranton; the iron ores around Scranton; Roaring Brook Falls, with Lehmann's very pretty sketch of the cascade facing page 338.

The coal beds of this basin are lettered A, B, C, &c. from below upward, according to M'Kinley's original classification. The series is shown in columnar fig. 262 at Scranton, and horizontal fig. 266. The Diamond Mine vertical section is given in figs. 269, 270, and its big coal in figs. 271–274; the same bed in the New York and Pennsylvania Company's slope in fig. 263; and coals B, D, F, in figs. 264, 265, 268.—M'Kinley's Red Shale (XI) section (fig. 275) deserves notice for its iron ore.

M'Kinley's long section across the basin through Scranton, (line 1, Plate V, end of volume,) extends from the top of the mountain south to the Providence road. Excellent as it was in that early day, it can now be greatly improved.

Chapter 3. Pittston and Mill Creek district (361–375); axes VI, VII, VIII, IX, X, XI, XII, XIII, XIV, XV, XVI (See fig. 278), and the corresponding basins from V to XVII are worked out by M'Kinley with great care, as they echelon across the valley. A general vertical section at Pittston is given in fig. 276. A construction of the underground by means of the slope of the Pennsylvania Co. at Port Griffith, and boring (fig. 278) is very instructive. The coal beds are traced from Wilcox's towards Pittston, and two coal bed sections are given (figs. 280, 281.) Finally a long section across the basin at Pittston is given on line 2, of Plate V, at the end of the volume.

Chapter 4. Wilkesbarre, Nanticoke and Shickshinny district, south side of the river (376–405); axes and basins XVII, XVIII, XIX, and their sub-divisions are described with particularity by M'Kinley, whose short cross-sections, figs. 286, 289, 291, 292, 293, 296 and 298, and long sections across the basin between Wilkesbarre and Pittston, and Mr. Rogers' very thorough section up through Solomon's gap, on lines 3, 4, 5, 6 and 7 of Plate V at the end of the volume—give a very fair view of the geology of this important coal field.

Its chief feature is, of course, the Mammoth or Baltimore bed, vertical cuts of which, at the old mine (fig. 283,) at Blackman's (fig. 287), at Landmesser's (fig. 288), are given; while Lehmann's colored interior of the mine, facing page 382, is a worthy representation of its magnificence, rivaling as it does the rock-temples of India. Lehmann's colored exterior of the mine (vignette to Volume II) is almost as fine; and yet these plates are but shrunk and insufficient reproductions of the much finer and larger originals. An exquisite sketch, after

Lehmann's picture oı the Nanticoke gap, faces page 399, and exhibits the curious coquetry of the Susquehanna in leaving the valley, after having entered ıt at Pittston, to flow between the Conglomerate ridge and the White Catskill (Shickshinny) mountain; then after a number of miles turning sharply to cut off its end, leaving a minute patch of its coal mounted on a pedestal of conglomerate, fractured curiously as shown in figs. 297, *a* and *b*.

The Coal Measures are exhibited in the tunnel sections, and by vertical sections (fig. 290) on Carey's run, and (fig. 294) at Hog-Back 4 miles west of Nanticoke.

One of the most curious crush-faults in any coal field is beautifully pictured in fig. 284, a section of anticlinal XIX, near the foot of the lower plane in Solomon's gap.

Chapter 6. Wyoming, Kingston and Plymouth district (405–412); with special descriptions of the Great Bed at Plymouth (figs. 303, 305, 306), Bennett's bed (fig. 307), Cooper's Orchard bed (fig. 308), and the Gould bed. Small section of other beds are given; a tall vertical column of all the beds in the measures (fig. 299); and interesting cut of the thinning away of the Great Bed (fig. 303), and two small cross-sections north of Plymouth (figs. 310, 311.) A large body of valuable knowledge, since obtained, awaits publication.

It ought to be known that besides numerous surface maps and underground maps of collieries by the Messrs. Harden, father and son, and other mining engineers, there is a fine large contour-line map of the whole Third Anthracite Coal Field published by Mr. Rothwell, which would be a most valuable contribution to the reports of the Second Geological Survey.

APPENDIX TO BOOK VIII.

This is Mr. Rogers' revision of the Pottsville or Southern Anthracite coal field, purporting to be carefully collected and verified notes of collieries, bringing down the knowledge of the field to 1858, the date of publication of the Final Report, (pages 413–442). These notes are arranged in an order from east to west: at Tuscarora, Middleport, Casca-William's valley, Silver Creek, Zachariah run, East Rosendale, Crow Hollow, Mill Creek, East and West Norwegian, West Branch, Minersville, Wolf Creek, W. W. Branch, Muddy Branch, Swatara, Middle

Creek, Donaldson, Raush Gap, Bear Gap; and in the Mine Hill valley. No illustrations are given, except one of a ground plan of the Otto White and Grey Ash collieries, fig. 312, (434); but these notes contain a body of valuable records of that date, and descriptions of discoveries in the course of operations. The most interesting relates to the splitting of the Mammoth Bed into three large widely separated beds, with separate local names, towards the western end of the field, and this forms a standard of comparison for similar phenomena in all coal regions, having a most important bearing upon the Theories of Coal.

BOOK VIII—*Continued* (448–465.)

This is a short and unsatisfactory chapter on the Broad Top Coal Basin, with a little wood-cut map of the north prong of East Broad Top coal basin (fig. 314); eight cuts of coal beds; a vertical column in Round Knob (fig. 322), and another on Sandy Run (fig. 324.) This interesting region has been elaborately surveyed by Lesley and Fulton, and the map after being revised and enlarged in 1875 and 1876, will be published in the Reports of Progress of the Second Survey.

On page 465 Mr. Rogers gives his short summary table of anthracite, semi-bituminous and bituminous coal areas, the total coal area in Pennsylvania being estimated at 12,655.77 square miles.

BOOK IX. *Bituminous Coal Fields.*

DIVISION I. *Vespertine (X) and Umbral (XI) series in the Bituminous Coal Region* (466.)—Their topographical relations.

Chapter 1. *Vespertine* (*X*) conglomerate; modifications of type and thickness (467.)

Chapter 2. *Umbral* (*XI*) red shale (468–473); geographical distribution; modifications of type; columnar sections at Lick run; Ralston; Conemaugh gap; Laurel Hill; Chestnut Ridge (figs. 326–330); 1,050 feet of rocks from the coal down to VIII are described in this last section of the fine gap at Blairsville. Lehmann's landscape of the plains of Western Pennsylvania, as seen through the Loyalhanna gap from the turnpike summit of Laurel Hill in Somerset county, faces page 468.

DIVISION II. *General View of Bituminous Coal Strata.*

Introductory chapter. Coal strata of Western Pennsylvania (474–479.) General vertical columnar section, compiled from all the counties west of the Allegheny and Monongahela rivers from Mercer to Greene. General description of the basins and axes.

SUB-DIVISION I. *North of the Conemaugh and Ohio.*

Chapter 1. *First basin and axis* (480.)

Chapter 2. *Second basin and axis* (482.)

Chapter 3. *Third basin and axis* (483.)

Chapter 4. *Fourth basin and axis* (485.)

Chapter 5. *Fifth basin and axis* (486.)

Chapter 6. *Sixth basin and final outcrop* (488.) Much of this description holds good; but the careful study of Clearfield county by Mr. Platt in 1874, has seriously modified the old classification of J. T. Hodge. The 1st and 2d basins of Lycoming become the 1st of Cambria, and therefore that the 3d basin of Potter is the 2d basin of Clearfield, &c. Also the 1st basin is sub-divided into two by short anticlinals along its centre-line, and that its two sub-divisions correspond to Hodges' 1st and 2d basins. It is too late to alter the numbering of the basins as a whole. In the reports of the Present Survey no change will be made, except in the Pine Creek and Lycoming Creek country, and there only by calling the 1st and 2d basins "first basin," the 3d "second," &c.

SUB-DIVISION II. *South of the Conemaugh and Ohio.*

Chapter 1. *First basin and sub-divisions* (494.)

Chapter 2. *Second or Ligonier basin* (497.)

Chapter 3. *Third basin and sub-divisions.* Upper or Greene County group (499–507.) In this chapter Mr. Rogers abandons the numbering of the basins south of the Conemaugh without good reason, for some of the anticlinals can be traced with great regularity across that artificial boundary. The remarkable anticlinals and synclinals worked out by Prof. Stevenson and Mr. White in Greene and Washington in 1875 were unknown in 1858, and would greatly change the descriptions in this chapter. The Pittsburg bed is well described, and two cuts of it given in fig. 332. Plate VII, at the end of the volume, gives one very long (6 lines) section of the coal measures along the Pennsylvania railroad from Altoona to Pittsburg.

SUB-DIVISION III. *Detailed description.*

Chapter 1. *First. Bituminous Basin* east of Pine creek (507–510); with four coal bed cuts of the one only workable coal bed of the region north of Williamsport.

Chapter 2. *Second Bituminous Basin* east of Pine creek (511–519); with columnar sections at Fall Creek, Gatiss' (Towanda basin), Ralston Old Mines, Astonville, Cartersville, and Otter's Run; and four cuts of the big bed of the Towanda basin; with a cross-section and landscape background of the valley which splits the Towanda mountain. Private surveys of the Towanda mountain will be published in 1876. The section at Ralston, (750 feet) given in the text and column on page 515 (fig. 345), shows the coal, Conglomerate (XII), Red shale (XI) and iron ore, Catskill rocks (X and IX) and top of the Chemung (VIII), with its iron ore near the bed of the Lycoming valley.

Chapter 3. *Third or Blossburg basin* (519–524); with seven small columnar sections of the coal measures; and one (fig. 350) of the whole series from the coal down to the Red Catskill (IX), 620 feet.

Chapter 4. *Fourth basin* east of Pine Creek (524.)

Chapter 5. *Fifth basin* east of the Genesee river (525).

Chapter 6. *Between Pine Creek and the Susquehanna West Branch* (526). These four chapters give nothing but the rude reconnoissance of the country made in 1841, when the country was a wilderness. In 1876 a careful survey of it will be commenced extending from Ridgeway, Warren and Renovo eastward to Blossburg and Ralston.

Chapter 7. *First basin* in Centre and Clearfield (527–533) with six little vertical sections of Hodge.

Chapter 8. *Second basin* in Centre and Clearfield (533–538) with three columnar sections of Hodge, and Lesley's two diagram sections with background to show the erosion of the Susquehanna valleys and ravines (figs. 364, 365.)

Chapter 9. *Third basin* in Clearfield and Jefferson (538–540) with one columnar section, and a cut of the Caledonia iron beds

Chapter 10. *Fourth basin* in Elk and Jefferson counties (540–545), with Lesley's eight columnar sections at Reynoldsville, &c. The country covered by the reconnoisance described in the above chapters, has been re-surveyed in 1874 by Mr. Franklin Platt, and described in his Report of Progress for that year, with 139 wood-cuts and ten maps and long sections.

Chapter 11. *Fifth and Sixth basins* in Potter, M'Kean, Elk and Forest counties (546–553), with five meagre columnar sections. This exceedingly important region was a wilderness when these notes were taken, more than thirty years ago. Railroads now traverse it, and a large population cultivates it; mines have been opened to supply the Buffalo market; and a thorough survey will be attempted in 1876, which should occupy at least two seasons. General Thos. L. Kane has spent twenty years in laborious and expensive researches, and has caused accurate local maps to be made; other explorers have also accomplished much; and a broad basis is thus laid for a new and complete survey.

Chapter 12. *Sixth basin* from M'Kean county south-westward, with columnar sections on the S. Branch of Kenzua creek and Tunamaguont; and nine cuts of the lowest coal bed and its relationship to the Conglomerate (XII). All this is very obscure and uncertain, and involves the difficult subject of the "Tionesta sandrock" and its supposed coal series. The Present Survey is studying the question *instrumentally* to reach reliable conclusions. Mr. Carll revolutionized our ideas of the stratigraphy in 1874, and has been laying a broad basis in 1875 for the settlement of the case in 1876. The investigation extends for more than a hundred miles, viz: from Beaver to Sharon, from Sharon to Meadville and Warren, and from Oil City to Ridgway; and nothing can be asserted with absolute confidence until the results along the whole line are collated and harmonized. Lesley's 1841 diagram section west of Smethport is given in fig. 386; and a subsequent rude vertical diagram of the hill-side rocks at Warren in fig. 392. Mr. Randall's section at Warren, in Carll's Report of 1874, improves upon it; but much has yet to be done before the relations of the conglomerates of the Catskill series, and the sandstones of the Chemung series, at and around Warren, are differentiated in a sure way.

Chapter 13. *Tionesta* (Sharon ?) coal series west of Franklin (561–564) with 5 columnar sections.—It is probable that the nomenclature to be adopted in future reports will distinguish three Conglomerates, upper, middle and lower; and that these will be called the *Pennsylvania Conglomerate* (XII), the *New York Conglomerate* (2d mountain sand, or Garland Rock,) and the *Berea Grit* (3d mountain sand.)

Chapter 14. *Bottom coals* of Mercer and Beaver and Lawrence counties (564–567) with eight columnar sections. A special survey in 1875 by Mr. Chance, and previous private surveys (See Wrigley's Report of Progress, 1874,) replace most of this chapter.

Chapter 15. *Clarion series*, in Jefferson and Clarion counties (567–572), with eight of Lesley's columnar sections of 1841.

Chapter 16. *Clarion series* west of the Allegheny river (572–576), with seven columnar sections of Hodge, M'Kinney, &c., and a cut (fig. 426) showing the gentle south dip of the Lower Coal Measures in the north-west corner of Lawrence county.

Chapter 17. *Freeport series* west of the Allegheny river (576–580), with three columnar sections, on Sugar creek, Allegheny river and Ohio river.

Chapter 18. *Freeport series* west of the Beaver river (580-583), with five columnar sections.

Chapter 19. *Fifth basin ; Red Bank ;* Jefferson county (583–586), with seven columnar sections, and Lesley's landscape background section of the valley of the Clarion at Armstrong's mill (fig. 435), showing the cap-plates of coal measures on the summits of the country. (See the contour-line maps at New Bethlehem, in Platt's Report of Progress for 1874.)

Chapter 20. *Third and Fourth basins ; Mahoning* and Cowanshannock creeks; Indiana and Armstrong counties (586–589); with eight columnar sections; fig. 443 showing the great Reynoldsville bed at Punxatawny.

Chapter 21. *Third basin* in Indiana county (589–591), with three columnar sections. This county and eastern Armstrong are almost unknown to geology. The district will be thoroughly surveyed in 1876.

Chapter 22. *Fourth and Fifth basins*, north of the Kiskaminitas; Kittanning and Freeport (592–599), with ten columnar sections carefully studied along the rivers; but the interior little known. A very remarkable wood-cut illustration of cross (oblique or current) bedding in Freeport sandstone at Leechburg is given (fig. 459.)

Chapter 23. *Third basin* (first sub-division) at Blairsville, Saltsburg, &c. (599–602), with four of Jackson's columnar sections. Fig. 466 gives 400 feet of measures on Chestnut Ridge.

Chapter 24. *Lower Coal Measures* along Chestnut Ridge, (Loyalhanna, Jacob's creek, Mount's creek, &c.) to the Maryland line (603–616), with three of Jackson's columns; five small cross-sections; two local-cut maps (figs. 471, 479) showing the outcrop of the Pittsburg (Connellsville) coal bed and iron ore; and three of Jackson's rude diagrams of extraordinary variations in the thickness of coal beds. It is a pity that these were not accurately portrayed to scale. As they stand they excite curiosity, but teach nothing reliable respecting the genesis of coal beds, nor respecting the possible water currents and changes of level, so often talked about in a reckless and ignorant manner. All such portraiture of abnormal exhibitions are mischievous, unless they be *mathematically* accurate in detail, and placed before the eye of the geologist *precisely* as they are in nature. The rough sketches, which field-workers too often indulge in, acquire authority when they are delicately engraved by artists (who are not of course geologists,) and do more harm than good, because they are accepted as real data for reasoning. The fancy of the bookmaker is tickled, but the earnest student is misled.

Chapter 25. *Pittsburg Coal Bed* and Upper Coal System facing Chestnut Ridge (616–628), with 29 sections of the bed, and 4 columnar sections including the coal beds above it, by Jackson.

Chapter 26. *Pittsburg series*, in the Second sub-division of the Third Basin (Jackson's Fourth Basin), with four cuts of the bed around Greensburg, and a columnar section at New Alexandria (fig. 514).

Chapter 27. *The Pittsburg coals* around Pittsburg (630–635), with a *collated* columnar section (fig. 515), showing 475 feet of measures from the Pittsburg bed down to the Upper Freeport bed made along the Allegheny river bank. Lehmann's landscape of Pittsburg and the union of the rivers to form the

Ohio, with the eddy-hill on which Allegheny College stands in the middle distance, faces page 630. Lehmann ascended Grant's Hill every day of the week in vain, and finally made his picture on sunday, when the vail of smoke was removed. The outline re-production here given poorly suggests the beauty of the original.

Chapter 28. *Pittsburg series* south of Greensburg in the third trough of the Third basin (636–650), with eight columnar sections of the rocks underlying the Pittsburg bed; eight cuts of the bed itself; thirteen, of the measures overlying it, including the Redbank, Sewickley and Waynesburg coal beds, and the great limestone. Some of these columns (of Jackson and M'-Kinley) give 500 feet of measures. Fig. 522 shows the subdivisions of the limestone. Fig. 528 is an instructive diagram section of its outcrop, creeping down hill and crushing out the slate, so as to permit the lowest layer of limestone to become the apparent roof of the coal bed In fig. 529 two diagrams show the cavernous erosion which has gone in the body of the limestone, and the subsequent interjection of ferruginous clay, Four cuts of the Waynesburg coal bed are given in figs. 545, 548. The whole of this chapter will be set aside by Prof. J. J. Stevenson's Report of Progress of 1875, in which his and Mr. White's thorough survey of Greene, Washington, South Butler and South Allegheny is described.

Chapter 29. *First and Second Basin;* Somerset and Fayette counties (650–665), with fifteen original columnar sections by Hodge and Lesley in 1840, by means of which they established the alphabetical system of the coal beds now so widely adopted. But it was never intended to apply beyond a limited region, and has done more harm than good. The Ohio Geologists have wisely repudiated it, but unwisely substitute an equally local and imperfect alphabetical nomenclature in its place. Its application to north-western Pennsylvania has turned out a partial failure; to say nothing of the absurd attempt to substitute it for the old and classic nomenclature of the Anthracite Collieries No permanent system of lettering our coal beds can be successfully adopted for many years; if, indeed, it be not impracticable from the very nature of the case, owing to the disappearance or coalessence of the coal beds when traced from one district into another. An earnest attempt by the writer in 1841, 1842, to accomplish a "harmony," merely in Western Pennsylvania, proved abortive and was abandoned. Where one man fails, however, another may succeed.—Lesley's little sketch section of the process by which Castleman's river is eroding its channel, is given in fig. 562; and a specimen of ancient erosion inside the coal measures, in fig. 563. In Stevenson's Report of Progress for 1875, underground (ancient) valleys of erosion, affecting large sections of the Barren Measures in Washington and Greene counties, will be described.—This chapter will be superseded by Mr. Platt's Report of 1875.

Chapter 30. *Sharon* (sub-conglomerate) *series* in southern Somerset and Fayette (665, 666.) This is a hint of the important expansion of the *lowest* coal measures through the Southern States. (See Stevenson's memoirs on Virginia coals.)

PART III. *Mesozoic Red Sandstone.*

INTRODUCTORY CHAPTER (667–671.)

Chapter 1. Detailed description (672–678.) Red shale and sandstone. Boundary in the Chester County lead region.

A colored geological map of the Lead and Copper mining district of Chester and Montgomery counties, along the line of junction of the Laurentian (?) Gneiss and the New Red, from Valley Forge, above Philadelphia, westward, occupies the space of two pages, and is bound in at page 674.

On this map are shown the mines, trap dykes, and patches of conglomerate at the base of the New Red; the termination of the North Valley Hill (Potsdam S. S. No. 1, ?) a little east of Valley creek; and the way in which the Auroral limestone laps round it, and slides under the New Red; a fine place to study this being in the ravine at Port Kennedy.

Chapter 2. *Lower Calcareous Conglomerate* west of the Susquehanna, along the south limit of the New Red (679, 680).

Chapter 3. *Upper Calcareous Conglomerate*, (Potomac marble) on the Delaware, Schuylkill, Susquehanna, &c. to the Maryland line (681–684); with a little section at Monroe on the Delaware (fig. 567); and a most remarkable chasm in the Auroral limestone (No. II) near Reading, filled with the Potomac marble breccia (fig. 568.)

This whole subject requires reconsideration, as the surveys of

1874 and 1875 in York and Adams counties show. The apparently unbroken dip of the New Red towards the north has been proved to be a deception by the mining operations at Dillsburg, and the English geologists have made a like discovery in their New Red country. The formation was accounted a barren field of research and neglected by the First Survey. It requires close and patient attention, throughout its whole extent, from the Present Survey; and no part of the State more needs to be accurately mapped in contour lines. Professor Frazer has found not only upthrows but a regular small anticlinal in the New Red, the southern border of which appears on all the cross-sections illustrating his Report of 1874.

Chapter 4. *Trap dykes* in the New Red (684–697); with Mr Rogers' cross-section sketch of a dyke through shales in a railroad cut near Gettysburg (fig. 569), showing the alteration of the shales near the planes of contact.—(See Frazer's Report of 1874.) The organic remains found in the New Red (692, 693). No mention is made by Mr. Rogers of Mr. C. M. Wheatley's large private collection at Phœnixville, which, if not soon purchased by the Legislature, or some public spirited citizen, will go, like Lesquereux's unrivalled collection of coal plants, to the Cambridge Museum, or to some other distant centre of science far outside the State to which it belongs. Mr. Wheatly secured the fossils thrown out by the miners when the Reading railroad tunnel was executed. Such an opportunity will not soon recur. Mr. Rogers discusses the theoretical age of our New Red, and decides against its reference to a Permian date, and in favor of calling it Triassic or Jurassic. To support this view he describes the New Red of the Connecticut river, and of the James, Dan and Deep rivers in the South (694–697.)

PART IV. *Igneous Rocks and Minerals, Veins and Ores.* The rarity of unstratified and igneous rocks in Pennsylvania; granitic dykes; greenstone (706).

Geological relations of the mineral veins near Phœnixville; Wheatley lode and mines, with its trap-dykes, its extent and condition in 1853, produce and branch lodes; Brookdale mine (707).

Granitic and trappean veins in the Gneiss north of the Chester valley; Knauertown iron and copper veins; Elizabeth copper mine; Steel's iron ore pits; Hopewell Furnace ore pits.

Warwick iron mines; Leighton iron mines (fig. 571); Warwick mine (fig. 572); Knauertown mine; others (708). Old copper mine in Lancaster county.

Catalogue of localities for mineral collectors (709).

Iron Ores of the Gneissic and Palæozoic Formations of Pennsylvania. Introductory remarks (712).

Chapter 1. *Magnetic and Chrome iron ores* (713–717). Metalliferous veins of the South mountains: Lehigh Hill; Durham; Mount Pleasant, Penn's mount; table of analyses.

Chapter 2. *Primal iron ores* (718–721):—Cornwall iron mine, with a section along the Lebanon valley (fig. 574), and across the three ore-hills, with a sketch map of the ore basin (fig. 575–577). This mine needs much further explanation. Other geologists take views opposed to those here stated.—Surface ores.—Table of analyses.

Chapter 3. *Auroral and Matinal iron ores* (722–728); 1. on the Lehigh; 2. between Allentown and Reading; with four cuts and long tables of analyses. This imperfect statement will be replaced by the Reports of Prof. Prime's Surveys of 1874, 1875 and 1876.

Chapter 4. *Fossil iron ores* of For. V (729–731), with a differentiated section at Hollidaysburg (fig. 583), showing also two small down-throw faults [of doubtful existence.] Table of analyses.

Chapter 5. *Iron ores of VI and VIII* (731, 732). Table of analyses.

Chapter 6. *Iron ores of VIII*, Devonian, (732, 733). Table of analyses.

Chapter 7. *Iron ore of XI.* Sub-carboniferous, (734). Table of analyses.

Chapter 8. *Iron ores of the Coal Measures*, (735–739). Prof. William B. Rogers' theory of the origin and accumulation of this ore. In this memoir the distinguished geologist of Virginia shows, first, that it abounds only in measures abounding in fossil plant forms;—secondly, chiefly near, or close to, coal beds;—thirdly, that it is absent in red formations, which are well known to be almost destitute of organic remains; but where the New Red, for example, holds coal beds, there also it holds beds of this iron ore;—fourthly, that in non ore-bearing formations, the iron is equally abundant, but has not been concentrated;—fifthly, that to explain the widely spreading ore beds by mineral springs or chemical precipitation is inadmissable;—sixthly, that all the sediments of the Coal Measure age were charged with sesquioxide of iron in a diffused state; so much of it as lay near organic matter was reduced to protoxide and concentrated in nodules or layers as protocarbonate of iron, by filtration and segregation, in the wet mud and sand. At the same time the mud lost its reddish color and assumed the characteristic grey tone of all coal measures. It is possible that the process has not even now wholly ceased. The carbonic acid set free from the organisms would dissolve the protocarbonate of iron diffused through porous sandstone and carry it down against the horizontal top-face of the next lower water-bearing clay rock, where it would in time by access of air alone, or in water, be turned again into limonite. Table of analyses (739, 740.)

Here properly the Final Report of 1858 ends. But its author added some important memoirs of his own on the geology of the United States and other countries, to show how important an influence the Survey of Pennsylvania had exerted upon the growth of the science everywhere.

Sketch of the Geology of the United States (741–775) with one illustration (fig. 584), a synoptical diagram of the European formations and their supposed equivalents in America.*

This memoir would make a very good text-book, if printed separately. The use of the fanciful names *Primal*, *Auroral*, *Matinal*, &c. in it throughout is a very slight blemish (if any), and the New York names are also given. The author's per-

*In this Mr. Rogers *seems* to adopt Murchison's error (which that geologist afterwards corrected) of making the Bala and Caradoc different rocks, instead of the same; but in fact he had the courage to side with Sedgewick in that unhappy controversy, as may be seen by the term Cambrian being applied to all below the Wenlock (our No V.) Mr. Rogers' Caradoc is a *sandstone*, and no doubt the May Hill sandstone (our Oneida, No. IV.)

sonal acquaintance with English geology, and the judicial character of his mind, fitted him for deciding properly many moot points which his scholarship enabled him to state clearly and forcibly in a condensed form. Being written previous to 1858, this memoir leaves out of view the geology of the Great West, which has been very recently explored.*

After a general introduction, the series of rocks commencing with the lowest (Primal, Potsdam), are successively described, always in the following order: *Description ; Thickness ; Geographical Distribution ; Fossils ; Equivalents abroad.*

It is remarkable that no allusion is made to the glacial theory of Agassiz in the latter pages of this memoir, which treat of the Pleistocene and Drift deposits, except in one sentence (page 775): "the new direction imparted is constantly such as a moving fluid *or plastic mass* would take in sweeping round or past the obstructing barriers."† The author maintains, but with a slightly relaxing grasp, his hold upon his own theory of an ocean translated across the continent by earthquake movements. Almost his last sentence (page 755) is: "The mere agitation or pulsating movement of the crust would suffice," &c.; and again: "the inundation would take the form of stupendous currents, the strewing power of which would be adequate," &c. Were Mr. Rogers happily living to re-write this memoir, under the influence of recent discoveries and discussions, his views would be greatly modified, and this memoir would be nearly all that is required for a text-book in our schools and colleges, if furnished with proper wood-cuts.‡

Mr. Lehmann's striking picture of the Spanish Hill near

---

* "The recent researches of Mr. Blake and Dr. Trask" (page 773) refers to the Mexican Boundary Survey and the very earliest Californian Survey, cotemporaneous with the first placer mining.

† For the latest, clearest and most concise statement of the whole subject of glacial erosion and transportation see J. G. Goodchild's paper (No. III) on Drift, in the London Geological Magazine, No. 129, Nov. 1874, pp. 496–510.

For a synopsis of Mr. Alfred Tylor's views see Geological Magazine No. 135, supplement, Sep. 1875; with curious wood-cut illustrations.

‡ It is very curious and instructive to observe how the old cataclysmic deluge of Rogers, as opposed to the glacial droppings of Agassiz, has been recently introduced into the Glacial Drift Theory by Dana, and the two woven in together. Dana's melting-ice flood is, however, a very different thing from Rogers' translated Arctic ocean flood.

Athens, Tioga county, Pa. is inserted, facing page 774, to illustrate the paragraph on the Northern Drift.

Conditions of the Physical Geography attending the production of the Palæozoic Strata of the United States (776–811.)

This admirable memoir has but one defect worth noticing, viz: its convulsive or catastrophic tone. The movements of the crust of the earth at the various epochs of nonconformability are represented as sudden and violent changes of the relative levels of land and sea; and Mr. Rogers shows throughout its pages how little sympathy he felt with the uniformitarianism of Lyell, and those extremely slow and reluctant changes of the earth's surface which all living geologists agree to presuppose in their hypotheses.

After preliminary observations, the author describes, in regular order from the oldest to the youngest, the depositions and disturbances of the Primal Period, and of the Auroral Period; the Auroral Blue limestone, &c.; the Palæontological break between the Auroral and Matinal series; the depositions and disturbances of the Matinal Period; the stupendous crust movement and revolution in the earth's inhabitants at the close of the Matinal Period;* disturbances of the Levant,—Surgent.—Scalent, &c. &c. up to the Umbral.

Here a new subject commences:—Of the Limits of the Appalachian Coal Strata;—Of the nature of Coal Strata;—Of the Phenomena connected with the Coal-seams;—Of the intimate Mechanical Structure of the Coal;—Of the Character of the Strata in immediate contact with the Coal-seams;—The Fire-clay floor;—Of the Roof of the Coal;—Of the direct contact

---

*Fig. 586, "Hasbrouck's quarry, Rondout, showing unconformity of Premeridian (VI) upon Matinal (II) strata," is probably a great and important mistake, as shown by the fault at Port Clinton on the Schuylkill, where the phenomenon is repeated in the same form, but with vertical Levant (IV,) and horizontal Matinal (III). This is certainly *not* a case of nonconformity, but a case of simple fracture and upthrow. All Mr. Rogers' conclusions, based upon his figure, should be entertained with grave doubts, if not with incredulity. In this fig. 586, we have the teaching of an old lesson, that the *sketcher* is a dangerous companion for the *geologist*. The upper part of the line of contact is cunningly drawn, so as to support the nonconformity hypothesis; whereas this upper limit of the exposure was probably so filled in with local fragments as to permit the sketch to be made *in any way* that best suited the notions of the observer. The downward crush of the horizontal rocks is of itself sufficient to prove the fact that it is a case of *upthrow*.

of Coal-beds and Marine Limestones;—Theory of the Origin of the Coal Strata; and in this the pulsating earthquake and tidal wave idea is most ingeniously used to account for the killing of each coal bed and its burial under a sandstone roof.*

The change from Bituminous to Anthracite is then explained in some paragraphs.—On Gradation in the proportion of volatile matter in the coal of the Appalachian basins; in which is shown the connection between a disturbed state of the crust and an anthracitic character of the coal.

The fossils of the coal are spoken of in general terms in the last paragraph, which serves as an introduction to Lesquereux's memoir further on.

A small separate memoir "Of the formative actions concerned in the production of the Palæozoic and other strata" (811–815) gives the author's views of the action of water in geology, and is illustrated by two wood-cuts (figs. 587 and 588) showing how Mr. Rogers conceived of the re-distribution of gravel and sand by the back-lash of a tidal wave.

The Fossils of Pennsylvania are described in two important memoirs, one by Mr. Rogers: On the Organic Remains of the Palæozoic strata of Pennsylvania (815–836), and the other by Mr. Leo Lesquereux on the Coal Plants (837–884).

Fossils of the Primal Strata: *Scolithus linearis* (fig. 589) and *Paradoxides spinosus* (fig. 590).

Auroral: *Leptæna incrassata* (591), *Atrypa plicifera* (592), *Murchisonia bicincta* (593), *Raphistoma staminea* (594), *Macluria magna* (585), *Orthoceras multicameratum* (596), and others not figured here, but figured in Hall's Palæontology of New York and elsewhere; about one hundred species in all, embracing plants, polyzoa, mollusca and crustaceans.

Matinal: II. 300 species, (200 restricted to the Trenton limestone,) nine-tenths of which differ from those of the Au-

* No allusion is made to the possible growth of coal beds over a continent perhaps at altitudes hundreds of feet above the ocean level; nor is this idea suggested in any of our text-books; although it may be used with much effect for explaining some of the latest discoveries respecting the bifurcation of coal-beds; and the origin of such deposits of coal as occur in the Missouri limestone.

roral limestone next below. Of these are figured *Chætetes lycoperdon* (597), *Strictopora acuta* (598), *Leptæna sericea* (599), *Leptæna alternata* (600), *Orthis testudinaria* (601), *Orthis pectinella* (602), *Lingula curta* (604), *Orbicula* (n. s. ? 603), *Ambonychia bellastriata* (605), *Pleurotomaria subconica* (606), *Bellerophon bilobatus* (607), *Orthoceras vertebrale* (609), *Conularia Trentonensis* (610), *Isotelus gigas* (611), and *Trinucleus concentricus* (612.)

Matinal Shale: III. Characteristic and common forms are given in figures: *Graptolithus priscus* (612), *Calymene Beckii* (613), *Agnostos lobatus* (614), *Lingula quadrata* (615), *Delthyris lynx* (616), *Avicula insueta* (617), *Modiolopsis modiolaris* (618), *Cyrtolithes ornatus* (619), *Pleurotomaria bilix* (620), *Ormoceras crebriseptum* (621). *Glyptocrinus decadactylus* (622); with a list of those prevailing in the back valleys. All these forms wholly disappear before reaching the next age of the—

Levant: IV. The principal fossil of these strata is the seaweed, *Arthrophycus Harlani* (623); with a small whorled shell *Bucania trilobatus* (624), and *Lingula cuneata.*

Surgent; V. Of about 120 species known in 1858, are figured: *Buthrotrephis gracilis?* (625), *Rusophycus bilobatus* (626), *Cornulites* (627), *Avicula subplana* (628), *Lingula oblonga* (629), *Leptæna depressa* (630), *Leptæna patenta* (631), *Atrypa congesta* (632), *Atrypa reticularis* (633), *Atrypa intermedia* (634), *Orthis elegantula* (635), *Orthoceras imbricatum* (636), *Calymene Clintoni* (637).

Scalent: V, Onondaga salt group and Water lime. *Beyrichia Pennsylvanica** and the little thorn-like *Teutaculites ornatus* (638) mark this horizon, which is rich in trilobites and other fossils. Mr. Rogers here figures (639) a spine of *Onchus* (?) *Deweii* (from Hall's Plate LXVI) and pronounces against it as a *fish* spine.

Premeridian: VI, Lower Helderberg; very fossiliferous; the most abundant forms being figured as *Atrypa peculiaris* (640), *Atrypa—?* (641), *Atrypa lævis* (642), *Delthyris macropleura* (643), *Spirifer sulcatus* (644), *Rhynconella nobilis* (645), *Pentamerus galeatus* (646), *Platyostoma* (647), *Strophomena punctulifera* (648).

* On page 834 Mr. Rogers prints notes by Mr. T. Rupert Jones, Curator of the London Geological Society's museum, on the nature of the *Beyrichia Maccoyiana, B. Pennsylvanica, Leperditia ovata, L. Gibbera v. scalaris,* and *L. Pennsylvanica,* which forms a large part of the mass of our No. V fossil ore; and gives figures of them (695-699) in various stages of their growth.

Meridian: VII, Oriskany. Undescribed *Calymene* and *Orbicula* near Moorestown, north of Montour's ridge, and at Frankstown. The characteristic and conspicuous forms are figured as *Meganteris ovoides* (649), *Spirifer arenosa* (650), and the most remarkable of all *Atrypa unguiformis* (651) or horse-print shell, by which this key rock of Pennsylvanian geology may always be recognized.

Post-meridian: VIII, Cauda-galli grit. Mr. Rogers gives no figure of the *fucoides cauda-galli*, or Cock's-tail sea-weed, and says that the shales which contain it (just over the Oriskany sandstone) scarcely appears in the State. But splendid specimens appear at Tyrone and elsewhere.

Cadent: VIII, Hamilton Lower black slate. The most common shell is *Atrypa limitaris* (652), and next, *Orbicula minuta* (653) and the delicate *Orthoceras sublatum* (654); *Cyathophyllum turbinatum* (655); a *Fenestella* (656); *Avicula decussata* (657), *Modiola compressa* (658), *Avicula flabella* (659), *Microdon bella-striata* (660), *Orthonota undulata* (661), other *Aviculas* (662, 663), *Strophomena crenistria* (665), *Strophomena demissa* (666), *Spirifer spiriferoides* (667), *Spirifer mucronata* (668) very abundant, *Delthyris medialis* (669), *Delthyris congesta* (670), *Atrypa aspera* (671), *Tropidolepis carinatus* (672), *Delthyris granulifera* (673), *Productus hirsutus* (674); the interesting coal plant *Lepidodendron primævis* (675), and with it, in the same slates, the delicate whorled shell *Goniatites interruptus?* (676). The most common trilobites are *Calymene bufo*, and *Homalonotus Dekaii.*

Vergent: VIII, Portage and Chemung. Of the scanty flora and fauna of this great mass in central Pennsylvania, are figured only one plant, a *Lepidodendron* (677); and *Avicula damnoniensis* (678), *Avicula* ——— (679), *Strophomena Chemungensis* (680), *Atrypa hystrix* (681), *Atrypa eximia* (682), and *Delthyris cuspidata* (683).

Ponent: IX, Lower Catskill. Only one figure (684) of a nameless *fern* is given in the text; but a wonderfully fine *leaf* is shown in Plate XXII, at the end of the volume.

Vespertine: X, Upper Catskill; and Umbral, XI red shale; describes Mr. Isaac Lea's discovery of four large lizard-like *foot prints* at Mount Carbon near Pottsville, and to the subsequent discovery of other smaller *foot prints* in the gap of the West

Branch 2 miles west of Pottsville; and gives a wood-cut (685) of corrugated impressions in the red shale; and another (686) of the supposed *track* of some shell-fish in the mud of that ancient shore. A beautiful *Lepidodendron* from X is given in Plate XXI, and a specimen of cock's-tail sea-weed. A magnificent plant "resembling a *Desmarestia*" from the red shale occupies Plate XXIII.

Coal Measures: In 1857 Mr. W. B. Rogers, Jr. discovered in coal slate near the mouth of the Ravensdale tunnel, casts of an *Avicula?* and of a *Tellenomya?* and these Mr. Rogers considers to be the first shells discovered in the anthracite measures of Pennsylvania. In the limestones of the Bituminous Fields some forms are abundant; on page 833 are figured a *Pentremites* (688), an *Avicula* (689), a *Nucula?* (690), a *Pleurotomaria* (691), *Nautilus decoratus* (692), a minute *Lingula* (693), and a large *Spirifer* (694.)

This memoir, then, gives wood-cuts of ninety (90) characteristic and commonly observed fossils (mostly shells) from most of the sub-divisions of the Palæozoic System in Pennsylvania, and would, if separately published, furnish a useful little hand-book to the citizens of the State. Its value to geologists studying the ancient flora and fauna is very small; but such students will necessarily resort to the large volumes of the New York Survey, and no publication of the fossils of Pennsylvania, however elaborately prepared, could set aside this necessity. A preliminary Report of Progress will be made by Mr. C. E. Hall, the Assistant in charge of the fossil collections of the Second Survey, showing the extent of such collections during the years 1874 and 1875. A few new specimens have been found.

It has been a subject of much speculation and enquiry what became of the large collections of fossils made by the Assistants of Mr. Rogers during the prosecution of the first survey, from 1835 to 1841. Three collections of the minerals remained in the State; but the fossils disappeared. It was supposed that Mr. Rogers took them to places in other States and in Great Britain where they could best be studied. He, himself, resided in Boston until he accepted a Chair of Natural History in the Glasgow University, where he died. But it is evident from the memoir above described that the fossils were never studied in any true sense of the term, for there are no *descriptions* of them in the Final Report; except the descriptions of the fossil plants of the Coal Measures by Mr. Lesquereax.

It now appears that Mr. Rogers took the State collection of fossils with him to Boston, and thence to Glasgow; whence they have been returned to Boston, and are now incorporated in the cabinet of the Boston Natural History Society; for, in the annual report of Prof. Alpheus Hyatt, Custodian of the Cabinet of the Boston Society of Natural History, dated May 5th, 1875, it is announced that the mineral collections of Henry D. and William B. Rogers has been included in the so called "American collection" of the society, revised by Mr. Crosby, and mounted by Miss Washburn.

Mr. Hyatt, after speaking of Mr. Crosby's arrangement, adds:—"This work includes the naming and mounting of the Henry D. and William B. Rogers' collections, *principally of fossils from Pennsylvania*, the Cleveland collection of Devonian specimens, and the formation of an Educational collection from the duplicates. The latter is progressing rapidly, and will soon be as complete as the society can afford to make it."

Mr. Hyatt goes on to say:—"The Rogers collections have suffered much damage from the loss of labels; this is particularly the case with the Henry D. Rogers collection, which was packed with the greatest carelessness by the parties entrusted with its transmission to America, after the death of Professor Rogers in Glasgow. The sudden illness of Mr. William B. Rogers, in the midst of his labors, unfortunately left his own collection also in great disorder. But it is a matter of sincere congratulation that the honored member of our society has so far recovered from his illness as to be again able to work with us. He has already reviewed the labels of a portion of his own collection, and expects to be able to continue his efforts until the whole of his own and his brother's collections have been revised. The south-east room in the basement was fitted up partly with the old cases which were removed from the former botanical room, and partly with the cabinets of the Rogers collections, and now serves as a general work room, as a lecture room and laboratory for the students of the Institute of Technology, and also as a storage room for the Rogers collections and the Educational collection."

It is probable that Prof. Hyatt has been misinformed respecting the alleged carelessness of the Glasgow packer; for the mineral collections of the Pennsylvania Survey were in the same condition 35 years ago; divorced from their field labels, and not all of them even ticketed with numbers; and worse than all, the catalogues had disappeared. The immense collections of that great and protracted survey, both mineral and organic, had therefore never any real scientific value. Nor was this in more than a small degree Mr. Rogers' fault. He had a large corps and a miserably small annual appropriation, to obtain which he had to make fight against a false sentiment of economy, and a universal ignorance and indifferance in relation to science. He had no proper head-quarters; no museum; no available paid assistance for museum work; and his surplus funds were spent in freight charges which were then high, while the service was slow and irregular. He always looked forward to a time when the field work should be ended, and he could set himself to an undisturbed study and arrangement of the collections. But that time never came. His corps was disbanded; his Final Report was to be written, and that without pay. The immense quarto which finally appeared should relieve his memory of any charge of want of interest in or effort to complete the work; for it was simply a physical impossibility that one man should do the work of many. His error lay in the attempt. There were a dozen men of science in Philadelphia ready and anxious to distribute among themselves the various tasks of studying the collections, mineral, vegetable and animal,—labeling, arranging and preparing descriptions of them for publication,—without other pay than the personal reputation to accrue therefrom. The Academy of Natural Sciences was then in its pristine activity, and its rooms were the natural and safe head-quarters of the State Geological Survey if Mr. Rogers chose to make them such, and to share the labors and honors of the great work with his fellows in science. The law compelled him to place a part of the collections at Pittsburg; there they were burnt up. The law obliged him to place duplicates at Harrisburg; there they were stolen, and finally sent to the Insane Asylum. The law prescribed the localization of the rest at Philadelphia; there he concentrated the best;

and all the fossils. Had he deposited these in the Museum of the Academy, if only during the six years of collection, they would probably have been in a state of systematic arrangement, and their descriptions published,—without trouble to himself, with profit to his reputation, and with benefit and pleasure to the experts who should volunteer, and to the world at large impatiently waiting for its appearance,—by the time the survey was finished. But he preferred to attempt impossibilities and to labor alone. The result was inevitable; nothing was published for many years; the collections remained unstudied and were in a great measure destroyed, and what remains is almost valueless. The minerals, without labels or catologue, in a garret in Philadelphia, are quite worthless. The fossils, also without labels and uncatalogued, are of no value except to the comparative zoologist, and in this sense may be said to have found a comfortable home, with a reasonable prospect of being of some use for instruction in the Museum of the B. N. H. Society in Boston. As it is now impossible to tell where they were found, they cannot afford us any help in our present efforts to characterize and identify our formations by their contained fossils.

The above story, thus unexpectedly brought into a clear light by the recent report of Prof. Hyatt, is fraught with instruction and warning. It shows how indispensable to the safety and utility of the collections of any great survey is a Museum Building, with a salaried curator, and a paid or volunteer service, to lable, arrange, study, describe and publish, as well as to give access to the public. Without some such permanent and convenient home for collections they *must* be wasted, scattered and destroyed; or even if fortunately preserved, they can be of no public utility.

Fossil Plants of the Coal Strata of Pennsylvania (835–884,) with XX Plates at the end of the volume.

In a short introduction Mr. Rogers states that in this memoir Mr. Lesquereux publishes 110 new species of fossil plants, by far the greater part of the specimens collected by himself; a few being first seen and studied in the rich cabinets of Mr. Clarkson, at Carbondale, in the Third Anthracite coal field, and of the Rev. Mr. Moore, at Greensburg, in the Third Bituminous coal basin.

These new species constitute about one-half of the total number of well-defined forms detected by him in the Coal Measures of Pennsylvania previous to 1857.* Of the 220 species then known, more than 100 had been found in Europe, and 50 more were extremely like European species; but all the actually new, or *American* species, also were closely related to those of Europe.

The species were more numerous and abundant in the Anthracite than in the Bituminous coal fields.

The white-ash lower coal measures are more sandy, gravelly, or disturbed in sedimentation than the red-ash upper measures, and their beds are more irregular; correspondingly local and variable is the vegetation as studied by Lesquereux. The lower beds abound also in large tree-plants, and the upper in small shrubby plants.

The aim of this distinguished botanist was to identify plants or groups of plants as *characteristic*, so that the miner could tell which of the beds he was working by noticing the plant-species in his roof slates. His success was, however, only partial; and he limits the applicability of this test in his memoir. But he considered the test good within certain limits. Sometimes, however, it failed signally; as for example, when he found the fossil-plants of the Salem vein so exactly like those of the Gate vein at Pottsville as to convince him that the two were one and the same bed; whereas, by the structural geology of the locality it was clearly proven that the Salem lies several hundred feet above the Gate.

For nearly twenty years since writing this memoir Mr. Lesquereux has continued to prosecute his researches on this important subject, in Kentucky and Arkansas under David Dale Owen, in Illinois under Mr. Worthen, in Indiana under Mr. Coxe, and in Kansas, and the Rocky mountains under Dr. Hayden; while Mr. Whitney has submitted to him the plants of

---

* In his essay on the botany of the coal measures, published in Hayden's Report of 1874, Mr. Lesquereux gives the number of species of *ferns alone*, known now, as 350; of which nearly one-half have been detected in the Carboniferous of the United States. This is an index of the progress of the science.

California, and large collections from other coal fields of America, tertiary and cretaceous, as well as carboniferous, have been placed at his disposal for study and publication. His writings are voluminous; his plates numerous and beautiful; and he stands naturally at the head of the fossil botanists of America. In 1857 the Second Geological Survey will publish his report on the Coal Plants of North America, with especial reference to the fossil botany of Pennsylvania, illustrated by himself with about sixty plates of characteristic species. Into this work he pours the treasures of knowledge of a long life of incessant labor in this field of his affections, and geologists will get a hand-book for field work, in the Coal Measure districts, as perfect as the most copious experience and highest ability can produce.

General Remarks on the Distribution of the coal-plants of Pennsylvania, and on the formation of the coal, (837–847) make the introduction to Mr. Lesquereux's

Description of the Fossil Plants found in the Anthracite and Bituminous Coal Measures of Pennsylvania (847–884.)

In his general remarks he discusses the subjects: which species are characteristic of the lower and upper measures; the botanical relationships of the anthracite and bituminous measures; the botanical resemblances of American and European coal measures; the completeness or incompleteness of the picture which the fossils afford us of the living vegetation of the coal age; the probable beginnings, growth and destruction of a coal bed; the analogy presented now by living peat bogs and cypress swamps; and the means at our command to comprehend the quality of the soil and climate of the world in the Coal Era, and in the Northern hemisphere.

In this descriptive part of the memoir Mr. Lesquereux divides the plants of the Coal Measures into classes, and describes the genera and species of each class separately:—

First Class.—*Fungineæ;* or mushrooms, only one of which, *Polyporites Bowmanni, Lindl*, had ever been detected. The small points on the stems of ferns may be mushrooms, but they may be accidental, for they occur abundantly on the surfaces of coal-slate, especially at Treverton, where the slates are sometimes full of holes (from the size of a needle's head to that of a pea) filled with a brown powder (like the spores of the *Spheriæ*), but under the microscope showing no structure.*

SECOND CLASS.—*Algæ ;* or sea-weeds. Not a single species of the true sea-weeds had then been found in the coal measures of Pennsylvania. Many species were known in the Sub-carboniferous, Devonian and Silurian rocks, and Mr. Lesquereux ascribed to their presence the petroleum. In 1863, however, he found fucoids in contact with a coal-bed, and others in coal measure rocks still higher up.—(See Report of Progress, J. 1874, third memoir: On Slippery Rock Creek.) Mr. Lesquereux divides the Algæ into:—

I. CHONDRITES: Sternberg.

1. *C. Antiquus.* Stern. (*Buthotrephis? flexuosa.* Hall)
2. *C. Fargionis.* Stern. (See Plate XVII, figs. 13 to 17.)

II. FUCOIDES. Harlan. (See *Pinnularia*, Plate XVII.)

1. *F. Alleghaniensis.* Harlan.
2. *F. Brogniarti.* Harlan.

THIRD CLASS.—*Muscites ;* or mosses, of which, although the coal-beds were largely composed of them, not a single species can be made out, so completely has the texture of the plant been destroyed. Experiments made with modern mosses show that they melt down into an indistinguishable mass when steeped for a long in water.†

FOURTH CLASS.—*Calamariæ ;*‡ or reeds, which are very abundant in all coal measure rocks, and are often seen as standing forests where the ocean breaks the coal measures down in vertical cliffs, as at Sydney, and on the Bay of Funda. They are divisable into:—

I. CALAMITES, Suck.; or REEDS

1. *C. decoratus*, Brongt,
2. *C. Suckovii*, Brongt.
3. *C. ramosus*, Brongt.
4. *C. cruciatus*, Stern.

---

* In the Australian and other coal fields resinous matters occur distributed in the rocks; and many coal beds are charged with vegetable pollen.

† See Hull on the Coal Fields of Great Britain.

‡ The *Asterophyllites* seems to be a part of this plant.—(See Plate I, figs. 1, 1a, 2, 3.)

5. *C. undulatus*, Brongt.
6. *C. Cistii*, Brongt.
7. *C. dubius*, Artis Antidil. Phitol.
8. *C. cannæformis*, Brongt.
9. *C. pachyderma*, Brongt.
10. *C. bistriatus*, Lesq. (Plate II, fig. 1.)
11. *C. disjunctus*, Lesq. (Plate II, fig. 5.)
12. *C. approximatus*, Brongt.

II. EQUISETES, Sternb.; or HORSE-TAILS.

1. E. stellifolius, Unger.*

III. ASTEROPHYLLITES, Brongt.

1. *A. gracilis*, Brongt.
2. *A. equisetiformis*, Brongt.
3. *A. foliosa*, Lindl. and Hutt.
4. *A. crassicaulis*, Lesq. (Plate I, fig. 1—1a.)
5. *A. ovalis*, Lesq. (Plate I, fig. 2.)
6. *A. sublævis*, Lesq. (Plate I, figs. 3, 9.)
7. *A. tuberculata*, Brongt.
8. *A. lanceolata*, Lesq. (Plate VII, fig. 32.)
9. *A. aperta*, Lesq. (Plate I, fig. 4.)
10. *A. Brardii*, Brongt.

IV. ANNULARIA, Brongt.†

1. *An. minuta*, Brongt.
2. *An. fertilis*, Sternb.
3. *An. longifolia*, Brongt.
4. *An. sphenophylloides*, Unger. (Plate I, 5, 5a.)

V. SPHENOPHYLLUM, Brongt.‡

1. *S. Schlotheimii*, Brongt.‖ (Plate I, 8, 8a, 8b.)
2. *S. emarginatum*, Brongt.
3. *S. filiculmis*, Lesq. (Plate I, fig. 6.)

---

* This is Harlan's *Equisitum stellifolium* and not a horse-tail, but an *Annularia*. What Harlan called *branches* are *leaves*. The plant which Harlan described, and which has caused geologists who could not visit the specimen to say that *horse-tails* grew in the Coal Era is a true coal plant, probably *Annularia fertilis* of Brogniart; but the real Equisetes came into existence *after* the Coal Era. Sternberg's *Equisetites mirabilis* is the stem of an unknown plant of which nothing more can be said.

† Lesquereux refuses Unger's classification of these with *Asterophyllites*.

‡ Ordinarily found with *Asterophyllites*, but although often seemingly never really belonging to the same plant.

‖ Entirely covering the slates of the South Salem vein at Pottsville.

4. *S. trifoliatum*, Lesq. (Plate I, fig. 7.)
5. *S. oblongifolium*, Unger.

Fifth Class.—*Filices;* or ferns, remains of which are far more abundant than of any other class in the coal measures; and the experimental maceration of living ferns in water for one or two years prove that they are the least destructable of plants. Their classification is difficult, because there are few reliable characteristic features. Göppert's new basis of classification—the form and position of the fructifications—is not good. The only available grounds of distinction between the species are the forms of the leaves and leaflets and the direction and ramification of the veins and veinlets.

I. NEUROPTERIDEÆ; froud primate, or bi-pinnate. Secondary veins, either rising from a medial nerve, vanishing above, or all emerging and branching from the base without any distinct medial nerve. No fruit ever found, although the plants are abundant.

1. *NOEGGERATHIA*, Sternb.
    1. *N. obliqua*, Göpp.
    2. *N. obtusa*, Lesq. (Plate I, fig. 11.)
    3. *N. minor*, Lesq. (Plate I, 10.)
    4. *N. Bockschiana*, Lesq.* (Plate III, 1, 1a, 1b, 1c, 1d.)
2. *CYCLOPTERIS*, Brongt.
    1. *C. flabellata*, Brongt.
    2. *C. fimbriata*, Lesq. (Plate IV, 17, 18.)
    3. *C. lacinata*, Lesq. (Plate XIX, 3.)
    4. *C. undans*, Lesq. (V, 3 to 7.)
    5. *C. elegans*, Lesq. (V, 4.)
    6. *C. trichomanoides*, Brongt.
    7. *C. hirsuta*, Lesq. (IV, 1 to 16.)
    8. *C. orbicularis*, Brongt.
    9. *C. Germari*, Göpp.
3. *NEUROPTERIS*, Brongt.
    1. *N. Rogersi*, Lesq. (VII, 2.)
    2. *N. hirsuta*. Lesq. (III, 6; IV, 1–16.)
    3. *N. Clarksoni*, Lesq. (VI, fig 1.)
    4. *N. fissa*, Lesq. (III, fig. 2.)

* In No. X, opposite Mauch Chunk. (IV, 19 to 22; V, 1, 2.)

5. *N. smilacifolia*, Sternb. (XX, 4.)
6. *N. plicata*, Sternb.
7. *N. flexuosa*, Brongt.
8. *N. Loschii*, Brongt. (XX, 3.)
9. *N. rotundifolia*, Brongt.
10. *N. tenuifolia*, Brongt.
11. *N. gigantea*, Brongt.
12. *N. Grangeri*, Brongt.
13. *N. Cistii*, Brongt.
14. *N. dèlicatula*, Lesq. (XX, 2.)
15. *N. Willersii*, Brongt. (III, 3.)
16. *N. gibbosa*, Lesq. (V, 3.)
17. *N. undans*, Lesq. (V, 1, 2.)
18. *N. crenulata*, Brongt.
19. *N. tenuinervis*, Lesq. (V, 7, 8.)
20. *N. dentata*, Lesq. (V, 9, 10.)
21. *N. Desorii*, Lesq. (V, 11, 12; XX, 5, 6, 7, 8.)
22. *N. heterophylla*, Brongt.
23. *N. minor*, Lesq. (III, 4.)
24. *N. rarinervis*, Bunb.
25. *N. Moorii*, Lesq. (XIX, 1.)
26. *N. adiantes*, Lesq. (XX, 1.)

4\. *ODONTOPTERIS*, Brongt.

1. *O. squamosa*, Lesq. (XIX, 2, 26.)
2. *O. Brardii*, Brongt.
3. *O. crenulata*, Brongt.
4. *O. Schlotheimii*, Brongt. (VII.)
5. *O. dubia*, Lesq.

5\. *DYCTIOPTERIS*, Gutb.

1. *D. Obliqua*, Bunb.* (VIII, 6.)

II. SPHENOPTERIDEÆ; frond bi-tripinnate or bi-tri-pinnatifid; pinnules sometimes entire, but mostly lobate, the lobes wedge-form at the base, dentate or diversely divided; nerves pinnate, with a primary nerve more or less distinct and flexuous; secondary nerves obliquely ascending, either simple in each lobe or division, or dichotomous, fureate at the apex; fructifications punctiform or marginal.

* South Salem vein and at Treverton.

1. *SPHENOPTERIS*, Brongt.
    1. *S. Darallia*, Göpp.
    2. *S. tenella*, Brongt.
    3. *S. Gravenhorstii*, Brongt.
    4. *S. Dubuissonii*, Brongt.
    5. *S. abbreviata*, Lesq. (IX, 1t, 1b.)
    6. *S. intermedia*, Lesq. (VIII, 8, 9, 9a.)
    7. *S. flagellaris*, Lesq. (XVIII, 1.)
    8 *S. plicata*, Lesq. (IX, 3.)
    9. *S. latifolia*, Brongt.
    10. *S. acuta*, Brongt.
    11. *S. glandulosa*, Lesq. (IX, 2.)
    12. *S. decipiens*, Lesq. (XVIII, 2.)
    13. *S. polyphylla*, Lind. and Hutt.
    14. *S. Newberryi*, Lesq. (IX, 4.)
    15. *S. Lesquereuxii*, Necob. (X, 1.)
    16. *S. squamosa*, Lesq. (X, 3.)
    17. *S. artemesiæfolia*, Brongt.
2. *HYMENOPHYLLITES*, Göpp.
    1. *H. furcatus?* Göpp
    2. *H. Hildreti*, Lesq. (IX, 5, 5a.)
    3. *H. capillaris*, Lesq. (IX, 6.)
3. *PACHYPHYLLUM*, Lesq.
    1. *P. fimbriatum*, Lesq. (VIII, 2.)
    2. *P. affine*, Lesq. (VIII, 1.)
    3. *P. hirsutum*, Lesq. (VIII, 3.)
    4. *P. laceratum*, Lesq.
    5. *P. lactuca*, Lesq. (VIII, 4, 5,)

III. PECOPTERIDEÆ; frond either simple, pinnate, bi-tripinnate, or bi-tripinnatifid; pinnules attached to the rachis by the whole of an equal or dilated base, ordinarily united to gether, and very seldom attenuated; medial nerve strongly marked; secondary nerves or nervules ordinarily perpendicular to the medial nerve, or diverging from it, simple, and rarely forking or dichotomous, bi-trifurcate. Fruit-dots marginal, either attached to nerves and lengthened, or punctiform.*

* All Göppert's species except *Asplenites nodosus*, characterized by form and position of fructification, are removed from this section by Lesquereux The genus is overcharged with heterogeneous species.

1. *ASPLENITES*, Göpp.
    1. *A. rubra*, Lesq.
2. *ALETHOPTERIS*, Sternb. and Göpp.
    1. *Al. lonchitidis*, Sternb.
    2. *Al. Pennsylvanica*, Lesq. (XI, 1, 2.)
    3. *Al. aquilina*, Göpp.
    4. *Al. urophylla*, Göpp.
    5. *Al. Serlii*, Göpp.
    6. *Al. marginata*, Göpp.
    7. *Al. distans*, Lesq. (XII, 2.)
    8. *Al. obscura*, Lesq. (I, 13, 14.)
    9. *Al. serrula*, Lesq. (XII, 1.)
    10. *Al. nervosa*, Göpp. (XVIII, 3.)
    11. *Al. lævis*, Lesq.
    12. *Al. muricata*, Göpp.
3. *CALLIPTERIS*, Brongt.
    1. *C. Sullivantii*, Lesq. (V, 13.)
4. *PECOPTERIS*, Brongt.
    1. *P. Cistii*, Brongt.
    2. *P. polymorpha*, Brongt.
    3. *P. distans*, Lesq. (XI, 3.)
    4. *P. velutina*, Lesq. (XII, 3.)
    5. *P. ovata*, Brongt.
    6. *P. notata*, Lesq. (XVIII,4.)
    7. *P. oreopteridis*, Brongt.
    8. *P. pusilla*, Lesq. (XI, 4.)
    9, 10. *P. arborescens*, Brongt.
    11. *P. cyathea*, Brongt.
    12. *P. arguta*, Brongt.
    13. *P. abbreviata*, Brongt.
    14. *P. unita*, Brongt.
    15. *P. concinna*, Lesq. (XI, 5.)
    16. *P. pennæformis*, Brongt.
    17. *P. plumosa*, Brongt.
    18. *P. Sillimanni*, Brongt.
    19. *P. Loschii*, Brongt.
    20. *P. decurrens*, Lesq. (XI, 5a.)
    21. *P. incompleta*. Lesq. (I, 12.)

Besides these, are ferns of undetermined affinities.

CREMAPTORIS, V. P. Schimper.

1. *C. Pennsylvanica*, Lesq. (III, 5.)

SCOLOPENDRITES, Lesq.

1. *S. grossedentata*, Lesq. (VIII, 7.)

Certain stems of trees are also uncertainly related to other fossil remains; of these are:—

I. CAULOPTERIS, Lind. and Hutt.

1. *C. punctata*, Lesq. (XIII, 1.)
2. *C. gigantea*, Lesq. (XIII, 2.)
3. *C. Cistii*, Brongt.

II. PSARONIUS, Corda.

III. DIPLOSTEGIUM, Corda.

1. *D. Brownianum*, Corda.

IV. STIGMARIA, Brongt.

1. *St. ficoides*, Brongt.*
2. *St. anabathra*, Corda.
3. *St. costata*, Lesq. (II, 3.)
4. *St. umbonata*, Lesq.
5. *St. irregularis*, Lesq. (II, 4.)
6. *St. radicans*, Lesq. (II, 2.)
7. *St. minuta*, Lesq. (XVI, 1, 2.)

V. SIGILLARIA, Brongt.

1. *Si. lepidodendrifolia*, Brongt.
2. *Si. sculpta*, Lesq. (XIII, 5.)
3. *Si. obliqua*, Brongt
4. *Si. fissa*, Lesq. (XIII, 4.)
5. *Si. dilatata*, Lesq. (XIII, 5.)
6. *St. Schimperi*, Lesq. (XIV, 1.)
7. *Si. stellata*, Lesq. (XIV, 2.)
8. *Si. Menardi*, Lesq.
9. *Si. Brardii*, Brongt.

* Lesquereux here corrects an old error. Sir W. E. Logan's discovery that this fossil characterizes the underclay of all coal beds has been accepted by all geologists. Lesquereux says that it is a mistake. He has found the fossil evenly distributed in the whole extent of the coal basin, but generally in greater abundance in the low beds of coal. Sometimes an astonishing abundance of it prevails in the *roof* slates, as well as in the floor-clay; in certain localities excluding all other forms of vegetation.

10. *Si. Defrancii*, Brongt.
11. *Si. Serlii*, Brongt.
12. *Si. tessellata*, Brongt.
13. *Si. elegans*, Brongt.
14. *Si. Broehantii*, Brongt.
15. *Si. alveolaria*, Brongt.
16. *Si. scutellata*, Brongt.
17. *Si. Sillimanni*, Brongt.
18. *Si. oculata*, Brongt.
19. *Si polita*, Lesq. (XIII, 3.)
20. *Si. dubia*, Lesq.
21. *Si. obovata*, Lesq. (XIV, 4.)
22. *Si. reniformis*, Brongt.
23. *Si. lævigata*, Brongt.
24. *Si. elongata*, Brongt.
25. *Si. rugosa*, Brongt.
26. *Si. alternans*, Lind. and Hutt.
27. *Si. catenulata*, Lind. and Hutt.
28. *Si. discoidea*, Lesq. (XIV, 5.)

VI. SYRINGODENDRON, Stern. and Brongt.

1. *Sy. pachyderma*, Brongt.
2. *Sy. cyclostegium*, Brongt.

VII. LEPIDODENDRON, Sternb.

1. *L. aculeatum*, Sternb.
2. *L. rugosum*, Sternb.
3. *L. crenatum*, Sternb.
4. *L. obovatum*, Sternb.
5. *L. modulatum*, Lesq. (XV, 1.)
6. *L. giganteum*, Lesq. (XV, 1.)
7. *L. vestitum*, Lesq. (XVI, 3.)
8. *L. conicum*, Lesq. (XV, 3.)
9. *L. oculatum*, Lesq. (XVI, 4.)
10. *L. distans*, Lesq. (XVI, 5.)
11. *L. rimosum*, Sternb.
12. *L. obtusum*, Lesq. (XVI, 6.)
13. *L. carinatum*, Lesq. (XV, 4.)
14. *L. clypeatum*, Lesq. (XV, 5; XVI, 7.)
15. *L. sigillarioides*, Lesq. (XV, 6.)
16. *L. Mieleckii*, Göpp.

VIII. ULODENDRON, Rhodes.

1. *U. majus*, Lind. and Hutt.
2. *U. Lindleyanum*, Sternb.

The fruits found in the coal measures are divided into the following genera and species:—

I. LEPIDOPHYLLUM, Brongt.

1. *L. acuminatum*, Lesq. (XVII, 2.)
2. *L. obtusum*, Lesq. (XVII, 3.)
3. *L. lanceolatum*, Brongt. (XVII, 1.)
4. *L. affine*, Lesq. (XVII, 5.)
5. *L. hastatum*, Lesq. (XVII, 7.)
6. *L. brevifolium*, Lesq. (XVII, 6.)
7. *L. plicatum*, Lesq. (XVII, 4.(
8. *L. lineari*, Brongt.

II. LEPIDOSTROBUS, Brongt.

1. *L. ornatus*, Lind. and Hutt.
2. *L. variabilis*, Lend. and Hutt.
3. *L. pinaster*, Lend. and Hutt.

III. BRACHYPHYLLUM, Brongt (?)

1. *B. obtusum*, Lesq. (XVII, 8.)

IV. CARDIOCARPON, Brongt.

1. *C. Trevortoni*, Lesq.
2. *C. plicatum*, Lesq. (XVI1, 9.)
3. *C. punctatum* (?)

V. TRIGONOCARPUM, Brongt.

1. *T. Schulzianum*, Göpp.
2. *T. Hildreti*, Lesq.
3. *T. oblongum*, Lind. and Hutt.

VI. RHABDOCARPOS, Göpp. and Berg.

1. *R. amygdalæformis*, Göpp. and Berg.
2. *R. venosus*, Lesq.

VII. CARPOLITHES, Sternb.

1. *C. fraxiniformis*, Göpp. and Berg.
2. *C. bifidus*, Lesq. (XVII, 10.)
3. *C. disjunctus*, Lesq. (XVII, 11.)
4. *C. acuminatus*, Sternb. (?)
5. *C. platimarginatus*, Lesq. (XVII, 12.)
6. *C. bicuspidatus*, Sternb.
7. *C. multistriatus* Sternb.

VIII.—C. umbonatus, Sternb. This is scarcely a fruit, and looks more like a nodule of iron ore than like an organic relic. The following also Lesquereux considered either wrong or wongly placed:—

I. CORDAITES.

1. *C. borassifolia*, Ung.

II. POACITES, Brongt.

III. CYPERITES, Lind. and Hutt.

IV. PINNULARIA, Lind. and Hutt.

1. *P. Calamitarum*, Lesq. (Plate I, fig. 9.)
2. *P. pinnata*, Lesq. (XVII, 18.)
3. *P. fucoides*, Lesq. (XVII, 19.)
4. *P. horizontalis*, Lesq. (XVII, 21.)
5. *P. capillacea*, Lind. and Hutt. (XVII, 22.)
6. *P. confervoides*, Lesq. (XVII, 20.)

At the close of this beautiful memoir Mr. Rogers remarks (page 878) that it was the result of Mr. Lesquereux's researches in 1852–1854 for the State of Pennsylvania,* and expresses what has always seemed to other geologists a wholly unjustifiable displeasure that the distinguished palæontologist, growing hopeless of seeing his descriptions of the rich new flora which he had discovered appear in print, should have secured his title to his numerous generic and specific names, first by including them in his report to the Kentucky Survey, and secondly by permitting the Pottsville Scientific Association to publish an enlarged and amended "Catalogue of Coal Plants of North America," read by him before that association in February, 1858.

This catalogue is reproduced on pages 878–883, and gives in the smallest space, and plainest form, all that was known of Coal Botany in American Geology in 1858.

On the laws of Structure of the more disturbed zones of the earth's crust (885–916.)

Classification of the several types of orographic structure visible

* "Under my direction, at the expense of the Commonwealth." It is evident that *such* work cannot be under the direction of another person, especially of one ignorant of the subject; and the amount and manner of *pay*, if told, would excite considerable astonishment.

in the Appalachians and other undulated mountain-cnains (917–941.)

In these two memoirs illustrated by sixty-six wood-cuts, chiefly diagrams (figs. 700–766), Mr. Rogers resumed all the physical laws of erosion, transportation, deposition, elevation and plication of rock masses discovered by the geologists of the survey, with certain theories of his own respecting the forces and agencies which might be supposed to have folded them up and swept them away. He lays down as a basis of reasoning the fluidity of the interior of the planet, the comparative thinness of the crust, the frequency and magnitude of earthquakes in former times, great waves in the subjacent ocean of lava crumpling and buttrassing up the crust, and great waves in the surface ocean of water sweeping over the uplifted folds and planing them down into low mountains and valleys. These theoretical speculations are now almost forgotten, but the collection and arrangement of the physical facts of the survey by which they were inspired are well displayed in these carefully written memoirs:—

1. The parallelism of the anticlinal and synclinal folds (886); 2. The groups into which they are thrown upon a map (887);—3. The forms of these waves; the normal form, leaning forward towards the north-west, as if the push came from the Atlantic side (889); their axis planes; their terminations (891);—4. Gradations in flexure (892);—5. Exceptional steepness towards the south-east (894), with three wood-cuts of sections on the Juniata, and a generalized section across the mountain belt (fig. 706);—6. Fractures and faults, crosswise; and whether our gaps have been caused by such transverse dislocations (895), with a wood-cut of the Delaware* Water Gap (fig. 707);—7. Fractures lengthwise; generally parallel to the axis planes; with six diagrams showing as many ways of faulting an anticlinal and synclinal curve (figs. 708–713); long faults; and Y-shaped faults;—8. Undulations and faults in Europe (899–902); the resemblance of the Jura to the Appalachians; the fan-shaped structure of the Alps, with a cross-section (fig. 714) looking very absurd since the Alpine studies of Favre, Studer,

* The instrumental survey of 1875 makes it almost certain that this gap was *not* caused by a fracture.

Hyrtl, &c. have been published ;—9. Slaty cleavage and its cause (902), with two diagrams showing its fan-structure on an anticlinal, and its alternate presence and absence in alternate strata ;—10. Foliation (904) ;—11. Prevailing theories of elevation (905), with a diagram (fig. 717) showing how Mr. Rogers supposed the rock-waves to be stiffened into form by injected lava ;—12. Views of geologists concerning cleavage and foliation (907) and his objections to them ;—13. Mr. Rogers' own theoretical views (911) with cuts of the New Hope and Gettysburg dykes (figs. 718–719) ;—14. General resumé, in 11 groups of theses, stating the whole subject in the fewest possible words (912–916.)

In the second treatise Mr. Rogers describes :—

1. The erosion of horizontal strata, into plateaux, simple escarpments, and terraced plains (917), with six diagrams (figs. 720–725) ;—2. The erosion of monoclinal strata, into ridged plains, wide valleys, crests, terraces, double crests, &c. (919), with twelve diagrams (figs. 726–737) ;—3. The erosion of anticlinal belts, into symmetrical mountains, canoe valleys, &c. (921) with three diagrams (figs. 740–742) ;—4. The erosion of synclinal belts into keel-shaped mountains, &c. (924), with seven diagrams (figs. 743–749) ;—5. Features of local erosion ; conical hills ; buttes ; terraces ; transverse ravines ; gaps ; notches, &c. (927–941), with sixteen diagrams (750–765), and a picture of a face of rock near Wilkesbarre covered with diluvial scratches.*

Coal Fields of the United States and British Provinces (942–968.) This is a sketch in four chapters of : 1. The Nova Scotia and New Brunswick fields ;—2. The great Appalachian field with a columnar diagram of the beds on the Cheat river in Virginia (fig. 767) ;—3. The Illinois field ; with two columns of measures at Henderson on the Ohio (figs. 768, 769) ;—4. The Iowa and Missouri field ;—V. The Missouri field.

Tables of analyses of all the best known Coals of North America (969–976). 1. Seven from the British provinces. 2. One of Rhode Island ; 55 of Pennsylvania ; 51 of Virginia ; 58 of Kentucky ; 1 of Indiana ; 36 of Southern Illinois ; 33 of Middle Illinois ; 38 of Northern Illinois ; 1 of Missouri.

* Omitted from page 774, Vol. II.

British Coal fields (977–987) with a table of the coal produce in 1834–'5,=64,351,079 tons from 2,613 collieries.

Coal; its composition, classification and varieties (988–997.) Four kinds of coal described.

Conditions in the Composition and Structure of Coal affecting its economic value (997–1006).—On page 1000 is given a table of relative density, composition, conditions of combustion, evaporating power, and consequent rank of 19 kinds of coal, foreign and domestic, anthracite and bituminous, to illustrate the statements made in this valuable paper, which deserves to be universally read and carefully considered, not only by students of geology and metallurgists, but by all who have dealings with coal as heating, smelting or steam generating agent.

Methods of searching for, opening and mining Coal, pursued in Pennsylvania (1007–1014.)

In this memoir the author describes first the *outward indications* of the presence of a coal bed (Lehmann's picture of the coal benches on Locust mountain); then the mode of finding it by *trial shafts* and the *creep* of its outcrop down hill (fig. 770), with Lehmann's very fine picture of the overturned outcrop near Newkirk; *boring* and *proving.*

Then follow the *methods of mining* coal, with a figure (771) of the *mouth of a gangway* in horizontal strata; Montelius' *open cut* gangway mouth, Mill creek (fig. 772); Lehmann's picture of the gangway entrance to bed 14, at the Nesquehoning mines, with side tracks and shutes; plan of *underground* workings on the Primrose bed, near Pottsville (fig. 773); *tunnels; slopes.*

Then a historical sketch of the introduction of coal as a domestic fuel, with Lehmann's picture of an anthracite breaker and slope, head-house &c. at Tuscarora.

American and European Coal fields forms a separate memoir (1015–1017) in which the areas of the coal fields of the world are tabulated; the total being 196,939 square miles for the United States alone. The comparative table on page 1017 is now superseded by more recent estimates. In a table of available coal quantities, Pennsylvania is given 316,400,000,000; and all North America 4,000,000,000,000 tons; which has been increased by the Rocky Mountain surveys of late years.

Comparative quantities of fuel in Coal lands and forest lands (1018);—*Dynamic value of coal* (1018);—*amount of coal mined* (1018);—*coal trade of Pennsylvania* (1018–1820);—from short separate chapters; illustrated with a double page copper-plate map of the coal fields of Pennsylvania by Lesley.

Statistics of the Iron trade (1020–1022) and a short account ot the Nickel mine in Lancaster county.

A few additional illustrations are given on pp. 1023, 1024; fig. 774, the snapped anticlinal of Nittany valley near Jacksonville, in which Mr. Rogers thinks the primal slates appear. This is doubtful, as will be seen by the measured sections of 1873, 1874 and 1876 when published.—Fig. 775, cross-section through the Kishacoquillas and Penn's valleys.—Fig. 776, cross-section in Bedford county.—Fig. 777, cross-section at Huntingdon.—and Fig. 778, cross-section, Whelpley's perspective cross-section from Tamaqua northward, showing the parallelism of the conglomerate ridges, &c.

A *Glossary* of terms (1025–1027.)

An *Index* (1028–1045) which, although copious, is both inadequate and imperfect. As the value of such a book must, for the majority of its readers, depend on the excellence of its index, it becomes an important question how to construct such an index; for while it seems needful that every geographical, mineral and personal name to be mentioned on any page should be found in the index, it is possible so to overload the index as to make it an intolerable labor to refer to such names. It is safe to say, however, that an index of only 18 pages is utterly insufficient for a book of 1,600 pages crowded throughout with names, to every one of which some reader will wish particularly to refer. In publishing the reports of the Second Geological Survey the greatest attention is bestowed upon the indexes. Dr. Genth's Report of 1874 has 180 pages of text and 26 pages of index. Mr. Wrigley's 107 of text and 14 of index. Mr. Carll's 108 of text and 19 of index. Mr. Prime's 66 of text and 7 of index. Mr. M'Creath's 93 of text and 10 of index,—total 554 pages of text and 76 pages of index, or one page of index for every 7½ pages of text. The Final Report of 1858 has but one page of index for 90 pages of text. Index-making is a special art, and for the purposes of science should be cultivated to the highest pitch. The labor of preparing a good index is great; but patience and perseverance will be rewarded by the gratitude of a thousand obliged readers. Discoverers lose the credit of their discoveries by burying them in the text of their report, without references in the index, and therefore out of sight of other investigators, who naturally turn to the index before they read a memoir, or in the press of their own occupations depend upon the indexes of books which they endeavor to collate. Personal quarrels would often be nipped in the bud if the candid opponent could remember whereabout in his antagonists' writings the obnoxious statements were to be found, that he might re-consider them carefully before he opposed them openly; but the index-maker has omitted them for want of patience or for want of system. Authors reverse their own statements, oppose their own conclusions, and forget or distort their own discoveries, after the lapse of years, because they have insufficiently indexed their own writings. And the public refuse to read many printed books because, as it is justly said, nobody knows how or where to find what one needs in them, for want of a perfect index.

The best way to make an index is to take some thousands of visiting cards, and write on them *separately* each geographical or other name which is encountered by the eye as it descends each page, with the folio (page) on which it occurs. When all are written the cards can be shuffled into an alphabetical order, and the index be written out for the printer. If besides a general alphabetical index, separate indexes of geographical names, personal names, mineral species, fossils, &c. &c. be required, the same cards may be shuffled again and again. Subjects of importance can be written out on some of the cards. If the same word occur on fifty different cards, but with different folios (page numbers), these folios can be all transferred to one card, thus diminishing the size of the pack to be shuffled before writing out the index for the printer. The only safe rule is to *omit nothing*, and not spare cards.

The Final Report of 1858 should have an index of at least 50 three-column pages to make it of easy reference. With all its defects, it is a book of rare beauty and value.

CHAPTER IV.

A sketch of the History of other State Geological Surveys in the United States, and of their relations to that of Pennsylvania.

A proper treatment of this subject would furnish for the student of the Science of Geology a useful text-book, and a guide book for the practical geologist. The literature of State Surveys has become so copious and cumbrous that earnest demands are made for a summary of facts. To supply this demand would task to their utmost the best talents in the closet and in the field. No such attempt is made here. But it is proper that he who surveys the geology of Pennsylvania should learn the beginnings, progress and general results of surveys in other States, and have some index to their publications; for he is not to presume that the whole circle of this science has been rounded out within the limits of even so great a State as this; nor that the types of stratification and structure revealed in it suffice to explain entirely the geology of even neighboring regions. The scale of our formations is large, but not complete; we must look beyond the limits of the State for its completion. Other States hold the keys to more than one secret of our State Geology. The reports of other geologists must supplement our own. This chapter is intended merely to direct attention to the number and value of such reports, and to indicate their authors, the dates at which they were written, and the circumstances under which they were published. Many of these printed records are old and effete; their authors dead; their statements doubtful or disproved. But a far greater number of them are reservoirs of facts well observed and carefully described, by able and zealous men, whose fame will last. It is not possible to do full justice to their genius for investigation and to their diligent and long continued labors, for lack of information of a personal kind; but it is not only possible but necessary to paint in some rude fashion, or draw in outline, the story of their doings, which is in fact the history of the growth and progress of Geology in America. There may be found in this an incentive for others who remember better, or have nearer means of information, to write a better treatise, in which the exaggerations, errors and omissions of this sketch shall be corrected.

A chronological order having been adopted in preceding chapters, the geological surveys of the United States follow each other here according to their dates of organization; and, to avoid an evident inconvenience, the work done in each State is brought down to 1875, whether the State has had but one, or several surveys.

NORTH CAROLINA.

North Carolina seems to have been the first or the United States to institute a Scientific Survey of its territory, in 1823. This honorable position was assigned to South Carolina by Mr. Vanuxem in his report on the geology of that State in 1826; but his successor, Mr. Tuomey, corrected the mistake when he published his report of 1848.

The first suggestion of a State Geological Survey seems to have been made by Judge Murphy as early as 1819.

In 1821 Prof. Olmsted wrote a letter on the subject, which was considered at a session of the Legislature, but no action was taken.

In 1823 a proposition of Prof. Olmsted to spend his vacations in North Carolina, was accepted by the Legislature, and an annual appropriation of $250, for four years, was made, and was afterwards continued two years longer, Dr. Mitchell, of Chapel Hill College, N. C., taking Prof. Olmsted's place as State Geologist, when the latter removed to Yale College. The reports of these six years explorations amounted, in all, to only 125 pages, 8°, and were published at intervals up to 1827. They related only to the eastern and middle portions of the State, and copies of them became very scarce at an early day.

In 1838 Governor Dudley urged a renewal of surveys; Governor Morehead did the same in 1844; and Governor Graham in 1846.

In 1851–'2, Dr. Ebenezer Emmons, one of the New York State Geologists was appointed State Geologist of North Carolina, with an annual appropriation of $5,000; and he continued in that office until his death in 1863. His son Dr. E. Emmons, Jr., became, in course of time, his assistant. Five reports were issued.

In 1852 Dr. Emmons published his first annual report on the Agriculture of the Lower Counties, and on the Coal Basin of Rockingham and Chatham counties.

In 1856, he published a Geological Report of the Midland counties; 351 pages, 8°; with 8 plates of fossils; sections across the State, and a colored map of the Deep River Coal field. In this very valuable volume, relating chiefly to the mines of the middle portion of the State, will be found his application to the southern field of his peculiar views of a "Taconic System," underlying the Palæozoic System, in eastern New York; views so vehemently opposed, so long discussed, and still unsettled.

In 1858, he published a Report on the Agriculture of the Eastern Counties, with descriptions of the fossils of the marl beds; 314 pages, 8°; with numerous wood-cuts in the text.

In 1860, he reported further on the Agriculture of North Carolina, and again on the Swamp Lands of the State. Dr. Curtis also published a report on the Woody Plants of the State. Dr. Curtis' reports on the Mammals and Reptiles of North Carolina remain unpublished; as also Mr. C. D. Smith's report on the Mountain District; and much geological matter relating to the whole State; as well as Dr. Emmons' proposed geological map of the State. During the civil war Dr. Emmons engaged in the business of manufacturing arms and ammunition for the Confederacy.

In 1864, after the death of Dr. Emmons, Prof. W. C. Kerr was appointed State Geologist; published his first annual report in 1865, 56 pages 8°; and has continued to publish similar annual reports to the present date. To that of 1871 Prof. Kerr added a treatise on the Mineralogy of North Carolina by Dr. F. A. Genth, which formed a valuable appendix to the volume, and has been separately printed.

SECOND GEOLOGICAL SURVEY OF PENNSYLVANIA.

Substance of a report to his Excellency JOHN F. HARTRANFT, Governor of Pennsylvania, President of the Board of Commissioners of the Second Geological Survey of Pennsylvania, dated Philadelphia December 31, 1874:

The Act of Legislature ordering the Survey, was passed May 14th, 1874.

The Board of Commissioners met for organization at Harrisburg, June 5th, 1874, at which meeting a Director of the Survey was chosen, and a general knowledge of the required work was obtained.

At the second meeting of the Board at Harrisburg, June 26th, 1874, the rules and regulations of the Board were adopted, salaries were fixed, appointments of assistants confirmed, the plan of the Director of the Survey approved, and appropriations made to meet the expenses of the ensuing quarter.

The plan of survey proposed and adopted after careful discussion and modification to suit the limited means placed by the Act of Legislature at the service of the Board, involved: 1. The occupation of five specially important and hitherto little studied districts of the State, requiring immediate attention. 2. The postponement for the present of work in the best known anthracite and bituminous coal regions. 3. The postponement for the present of systematic study of fossil forms, on a large scale. 4. The establishment of a special laboratory at Harrisburg for the analysis of irons, steels, iron ores and other blast-furnace stock. 5. A special report, chiefly economical, on petroleum. 6. A special report on the mineralogy of the State, as at present known. 7. The publication of the season's work early in the winter and spring, in a series of separate, portable and cheap volumes, so as to make them practically useful to the largest number of persons, each volume containing the illustrations of its own text, in the shape of maps and wood-cuts, and printed from stereotype plates for future use; and finally, 8. The exhibition of type specimens of the collections of the survey in a cabinet, in the rooms of the Board, at Harrisburg.

From among about one hundred and twenty applicants for service in the survey the following gentlemen were chosen to take charge of the five districts in which only the Board felt authorized to commence work at first, in view of the fact, that after the purchase of all the instruments required and the equipments of the Iron and Steel Laboratory and Cabinet Rooms at Harrisburg, there would remain of the first annual appropriation of $35,000, say $30,000 for the actual business of the survey:

Mr. Andrew S. McCreath, of Baldwin, assistant in charge of the Iron and Steel Laboratory, at Harrisburg.

Prof. Frederick Prime, Jr., of Easton, assistant in charge of the Lehigh district.

Prof. Persifer Frazer, Jr., of Philadelphia, assistant in charge of the York and Adams district.

Mr. John H. Dewees, of Shamoken, assistant in charge of the Juniata district, for the special study of the Fossil Ore belts.

Mr. Flanklin Platt, of Philadelphia, assistant in charge of the Clearfield and Jefferson bituminous coal district.

Mr. John F. Carll, of Pleasantville, assistant in charge of the Oil Districts of Western Pennsylvania.

Dr. F. A. Genth, of the University of Pennsylvania, Chem ist and Mineralogist.

Mr. Henry E. Wrigley, of Titusville, to prepare a special economic report on the petroleum of Pennsylvania.

Mr. Edward B. Harden, of Philadelphia, Draughtsman.

These gentlemen commenced work at various dates in the months of July and August, and have been actively employed to the present time. The exceptionally genial and open autumn weather of 1874 has permitted field-work to be prosecuted into December, and Mr. Dewees is still in the field.

The field parties were made up by the appointment of the following named gentlemen, all of them experienced Assistant Civil Engineers, trained on the various railroad lines of our State to the use of instruments, and good daughtsmen accustomed to reduce their field-notes to the form of the required maps, profiles, cross-sections, vertical sections, &c., and of their skill and diligence I cannot speak in terms too high:

Mr Ellis Clark, of Ambler, Pa., assistant to Prof. Prime; Lehigh district.

Mr. Ambrose E. Lehmann, of Lebanon, Pa., assistant to Prof. Frazer; York and Adams district.

Mr. Charles E. Billin, of Philadelphia, assistant to Mr. Dewees; Juniata district.

Mr. Richard H. Sanders, of Philadelphia, assistant to Mr. Platt; Clearfield and Jefferson district.

Mr. F. A. Hatch, of Johnstown, Pa., assistant to Mr. Carll; Venango and Armstrong district.

Mr. Frederick Watts Foreman, of Harrisburg, was appointed to act as the clerk of Mr. Pearse, the Secretary of the Board of Commissioners, to have charge of the rooms of the Board, in Harrisburg, and to assist Mr. McCreath in the Laboratory until the issue of the Reports of Progress should require his attention.

To render the various parties efficient, the following gentlemen were appointed additional aids, or served as volunteers, throughout the season:

Mr. Joseph R. Shimer, of Lafayette College, Easton, with Prof. Prime.

Mr. Charles H. Allen, of Harrisburg, aid to Prof. Frazer, and afterwards in the Museum at Harrisburg.

Mr. J. W. Edwards, volunteer aid to Prof. Frazer, and afterwards in the Lehigh district.

Mr. C. A. Ashburner, of the University of Pennsylvania, aid to Mr. Dewees.

Mr. G. H. Christian, Jr., of the University of Pennsylvania, aid to Mr. Dewees.

Mr. Arthur Hale, of Williams College, Mass., volunteer aid to Mr. Dewees.

Mr. H. J. Fagen, of the University of Pennsylvania, aid to Mr. Platt.

Mr. C. A. Young, of the University of Pennsylvania, aid to Mr. Platt.

Mr. Henry M. Chance, of the University of Pennsylvania,

volunteer assistant for making special surveys of the Delaware, Lehigh and Schuylkill Water Gaps.

The equipment of each field party consisted of the following instruments, manufactured by Heller & Brightly of Philadelphia, viz: one transit with vertical circle and cross hairs, for stadia measurement of distances, with telemeter rod; one surveyors' compass, with ball and socket on light tripod; one odometer, with ten foot wheel, copper tyre and Jacob staff, to which the compass was transferable; two hand (Locke) levels; two clinometers; steel chain, table instruments, note books, &c.

As reliable aneroid barometers of sufficient delicacy could neither be manufactured in this country nor imported from Europe by the ordinary channels of trade, and as the use of such instruments is indispensable for doing the best modern geological work, I visited myself, by permission of the Board, the factories of instruments of precision in Europe, sailing from Philadelphia the 13th of July, and returning to Philadelphia the 2d of September; having organized the field parties and seen them at work before my departure.

During the twenty-four days which I spent in Europe, I visited instrument makers in Rotterdam, Brussels, Neufchatel, Zurich, Paris and London. My conviction is that no aneroid barometer of the ordinary well-known patterns to be found in the shops ean be considered a useful instrument in the special work of those field geologists who insist upon obtaining the best practical results. I have used most of these forms, and have found them obnoxious to the following objections: 1. Their scale is so minute that the eyes are strained to read less than 40 or 50 foot variations of height; 2. Their action is irregular, owing to imperfect workmanship in the capsules and gearing, and especially to a back-lash in the joints; 3. Their asserted "compensation" for varying temperature is a delusion. One of the most carefully constructed London aneroids of high price, and marked "compensated," was recently sent to me for acceptance as a perfect instrument; but on repeated trials with a temperature-range between 35° and 95° F., it alternately lost and gained 125 feet of height-reading.

Twenty years ago I devised a form of aneroid to overcome these difficulties, and was fortunate enough to find in the distinguished balance-maker of Europe, Mr. Becker, one who could make it for me. To his consummate mechanical genius and perfect workmanship I owe an instrument which, after years of practical experience with it on many surveys, I can pronounce nearly perfect, as a geologist's aneroid. It has twelve vacuum-boxes, mounted in two piles of six each, the piles surrounded by a compensation yoke, and the index driven by a T rachet. The dial plate gives a circuit of 14 inches. The index travels 7 inches for one of the mercurial column, and can be adjusted to travel 7 inches for 1,000 feet of elevation within any section of the mercurial range. A set screw, placed beneath, carries back the index; so that the 2,000 foot circuit may be repeated eight times, if mountains are to be ascended not exceeding a height of 16,000 feet. My object was, however, to obtain a large visible motion of the index in a country like Pennsylvania, where the range from sea-level to the highest mountain summits does not exceed 3,500 feet. Heights of 2½ feet can be easily read at a glance. The compensation for thermometer is very good, and the instrument is very light. Ordinary aneroid barometers may satisfy explorers and travelers, who wish to record considerable variations of altitude by observations at considerable intervals of time. But an accurate field-geologist must read his aneroid *at every station, every few minutes*, and must distinguish variations of altitude amounting to only a few feet. The accuracy of his geological deductions will always depend on the truthfulness of his topographical delineations, and this again will depend on the number and reliability of his aneroid observations.

The best makers in Europe told me candidly that they were unable to reproduce Mr. Becker's aneroid. Elliott and Bros., Charing Cross, London, took drawings and measurements of its interior, and promised to send me a *fac-simile* for trial. Mr. Becker had originally made five of these instruments, two of which I own. A third was sent to the late Chief Engineer of the Grand Duchy of Baden, Mr. Eisenlohr, who reported enthusiastically upon its behavior at the Swiss Congress in Geneva in 1858 (?). A fourth was possessed by Mr. Thomas Blackwell, Managing Director of the Grand Trunk Railroad, until his death. The fifth was carried by a German traveler across South America, and has also disappeared. I have a hope that Mr. Becker may consent to manufacture more of these instruments for the use of the survey; but he has thus far refused, because, owing to the dearth of skilled workmen not fully employed in other equally difficult tasks, he will be obliged to undergo the labor himself, at an age when he desires leasure to realize his conceptions of a still more accurate aneroid than the one he made for me, but of a form not suited, in my opinion, to ordinary field work. These personal details may seem out of place in a report like this. But I feel sure that my fellow geologists, and all accomplished and ambitious civil engineers will be gratified to learn the history of an attempt to place at the disposal of their craft a piece of apparatus for which there has been persistent, growing and impatient inquiry.

Feeling that *some* good form of aneroid must be obtained for the survey, I ordered a dozen from M. Goldschmidt of Zurich, after testing specimens in his shop. This instrument is coming into use among civil engineers. It is a cylinder about 4 inches high, on the top surface of which are described several concentric scales, traversed by an index, moved by a button. By turning this button (and the index with it) the observer, looking through an attached microscope into the interior of the box, brings a needle point to a datum line. An observed number on a small scale (visible in the field of the microscope) informs him, then, by which one of the several scales on the lid of the box he is to read (with the large index attached to the button) *the height of his station of observation above tide, in metres.* Adjustments to half a metre may be made through the microscope.

For want of aneroid barometers the work of the survey has been greatly retarded. I feel sure that if two reliable aneroids could have been given to each field party, at least twice as much ground would have been gone over, and twice as large a report would now be in process of publication. A few of the shop aneroids were tried, and soon laid aside as useless, by the gentlemen in charge of the different districts, and were only used in exceptional parts of their work.

The method of survey pursued by all the field parties was

essentially the same; although the widely different characters of the districts occupied caused great differences in the areas gone over, and in the results obtained; as will appear by the published reports, and need not be specified here.

The method of survey adopted was as simple as a long experience of field work dictated to be possible.

Base lines in each district were run and leveled along selected main roads and the topography carefully studied; plotting and contouring being done in the note-book in the field. From these note-book plottings the maps have been made, both by re-plottings and by transfer.

The note-books of the survey are all alike, one page being cross-barred for plotting; blue lines an inch apart; red lines ¼ inch apart. The top of the page is always taken as north. The last station before a line runs off one page is carefully located in the same relative position within some square inch on the next page. Consequently transfer of plotted lines from note-book to sheet plotting paper is easy and continuous. The opposite plain-ruled page holds the notes and sketches. Each book is numbered, and has an index of lines run written on its front cover, and a sketch-map of connected lines on its back cover. All stations in one district are numbered continuously from 0 to 10,000, to prevent any two stations from having the same number. Stations numbered in red; level numbers and contour-lines in blue; houses, fences, &c. in black. A permanent original record of all the work done during the survey will thus be preserved in a shape convenient for reference. Economy in note-books is disastrous to any large survey.

A universal scale was adopted for the whole survey, of 400 feet to the inch in the original plottings, and 1,600 feet to the inch in the reductions. Local maps and sections will be published thus reduced to a scale of 1,600. District maps will be still further reduced by the camera to 3,200 or some still smaller scale. The necessity for a rigid adherence to a common scale will be appreciated by the geologists of the survey when they come to compare together their own numerous sections and those of each other, along the same belts of rock. It will make the publications more useful to experts and the public, for whose benefit the survey is intended.

While the aids carried on the instrumental work, the assistant in charge of the district studied its geology, made reconnoissances in advance, visited all the mines and important exposures, and determined the direction in which the field work of his party should move forward, and the amount of such work which was necessary or possible. Lines of outcrop were traced and specimens collected and forwarded to headquarters for analysis. Transfers of the plottings to sheets were made in bad weather, or in the evenings, and forwarded to Philadelphia for reduction in the office. Railway surveys were obtained by Mr. Harden, copied on the survey scale, and sent to the parties in the field for their use. He also provided all materials required by the field parties, and kept records and accounts ot what was done and expended. To his indefatigable diligence, attention and ability, is largely due the smooth working of so complicated a machine as a State survey necessarily is.

Citizens of the Commonwealth have exhibited much interest in the operations of the parties in the field, and furnished information of great value. Unpublished reports and records ot mines, borings, minerals and fossils have been freely tendered for publication. A great deal of the private geological work of recent years is at our command, and will eventually be used.

Such materials for knowledge are sure to be evoked by a State survey; and an important part of the business of the gentlemen in charge of districts has been to secure, examine and improve these materials for their reports.

But the main intention of the Second Geological Survey of Pennsylvania goes much beyond the collection of scattered items of unpublished information. The records of the First Geological Survey afford a thousand times more real information about the geology of the State than can be thus gathered up. The accuracy of most of the statements made by the geologists whose reports are condensed and consolidated in the final report of Prof. H. D. Rogers, published in 1858, is a matter of frequent remark and admiration with the assistants now in the field. Yet the first survey was essentially a reconnoissance. Those engaged in it thirty years ago worked chiefly without instuments of precision and under the greatest inconvenience. Their views were broad, their isolated observations numerous and exact, but their districts never were *accurately surveyed* by them, nor could be. The second survey is intended to supply this lack; to take up their work where they left off; to reduce their general statements to precision; to measure, where they could only estimate; to define, what they could only indicate; to demonstrate what they could see to be true, but which they could not prove and show in all its truth. Geological field work now, and geological field work forty years ago, differ as widely as the present methods of surveying an anthracite coal property differ from one of our old-fashioned land warrant surveys. And I trust that the reports of 1874, now in press, will make both the manner and the necessity of this difference so manifest to everybody, that none will be disposed to cavil at the quantity of work accomplished, or the scrupulous attention which has been expended upon its details.

The following reports are either already in the printer's hands, or nearly ready for publication:

A. My Report to the Board of Commissioners on the work of 1874, followed by a history of Geological Exploration in Pennsylvania from the beginning.

The object of this statement is to show what a Geological State Survey really means; what is its public necessity; the nature of the problems is solves; how it works; what are the difficulties it encounters; and its methods of overcoming them; its necessarily slow and laborious progress; the imperfection of its results; the immense accumulation of facts; its power to stimulate the intellect of the State; to sweep away costly superstitions respecting the mineral resources of the Commonwealth; and to increase the physical powers of society; what discoveries have been made since 1835, and what questions of geological importance remain unanswered.

This statement might be made in a more didactic form. But when sketched as a history, it gains in power over the imagination more than it loses in systematic arrangement. In the course of forty years every point of geological science has been in turn, and many times, touched by the geologists of Pennsylvania; and citizens of the Commonwealth ought to have exhibited to them the manly struggle of these past forty years—a contest between our youthful, growing, strengthening and maturing science and the powers of darkness of the Underworld, the stubborn instinct for concealment in Nature, and the prejudices and falsifications of half-educated men. If this strife be well painted, the future consequences of a victory of light over darkness, of science over ignorance, in relation to the Geology of our State, will be made clear to every one. There will be no longer need to plead or apologize for a State Geological Survey.

B. A Report on the Mineralogy of Pennsylvania, by Dr. F. A. Genth.

Dr. Genth gives the character of every kind of mineral known to be found in Pennsylvania, with references to their geographical situations, and to the public and private museums in which specimens of them may be studied. This little volume of about 220 pages will be a valuable text-book for the Public Schools, as well as a guide for the assistants on the survey.

C. A Report of Progress in York and Adams Counties, by Prof. Persifor Frazer, Jr.

A map of the district, by Mr. Lehmann, and profile sections illustrate this report. The map shows all the lines run and leveled to connect the ore banks, which are seen to lie in belts.

D. A Report of Progress in Lehigh County, by Prof. Frederick Prime, Jr.

A map of the district between Alburtis and Allentown, elaborately contoured by Mr. Clarke, exhibits the arrangement of about a hundred iron ore mines in three belts.

E. A Report of Progress in the Siluro-Cambrian Limestone Valleys of Middle Pennsylvania, by Mr. Franklin Platt.

F. A Report of Progress in the Juniata District, by Mr. John H. Dewees.

A map of the southern flank of Jack's Mountain from Logan Gap to Jack's Narrows, elaborately contoured by Mr. Billin, and—

Five profile sections drawn by Mr. Ashburner, illustrate this Report, which not only relates the outcrops and mine openings of the fossil ore of the mountain, and the brown hematite bed of McVeytown, but minutely defines the Silurian and Devonian formations from the Trenton up to the Hamilton. The map of the mountain is intended to show how all the mountain region of Pennsylvania *might be* and *ought to be* surveyed, if the Legislature should provide the necessary funds for such a work.

G. No work has been done in the anthracite region.

H. A Report of Progress in the Bituminous Coal Region of Clearfield and Jefferson counties, by Mr. Franklin Platt..

Maps of the Osceola region; of the Snowshoe basin; of the Clearfield basin; and of the Reynoldsville basin, in contours; with local vertical sections, about a hundred in number, of individual coal beds, and columns of coal measures; all drawn by Mr. Sanders; a map of the Red Bank cannel coal field; and a map of the bituminous coal adjoining; by Mr. Blandy, presented to the survey by Mr. John Wilson, will illustrate this report.

A report of a recent survey of the Johnstown coal field in Cambria county; and a map of the same by Mr. John Fulton, of Saxton, Blair county, will be published, by permission.

A section of the sub-carboniferous rocks of the Allegheny Mountain, underlying Mr. Platt's district, by Messrs. Young and Fagen, will accompany Mr. Platt s report.

I. A Report of Progress in the Venango Oil Region, by Mr. John F. Carll, of Pleasantville.

A map of lines run and well-mouths leveled between Oil Creek and the Allegheny River; with sections in two directions across the district, showing the number and slope of the oil-bearing and mountain-sand rocks; a vertical column of rocks below the coal measures, showing the horizons of fossils, &c., and a general map of the country, by Mr. Carll and Mr. Hatch, will illustrate this report.

A Report of his observations around Warren has been kindly volunteered by Mr. F. A. Randall, of Warren.

J. A Special Report on Petroleum, by Mr. Henry E. Wrigley, of Titusville, Venango county.

A small map of the Oil Regions of the Middle States and

Canada; a large map of the West Pennsylvania and West Virginia Oil Regions, with lines of geographical limitation between gas wells, light-oil wells, and heavy-oil wells; a profile section from Lake Erie to Butler, showing the general slope, under ground; a chart, showing the construction of derricks, seed-bags, bull-wheels, boring-tools, &c.; and a chart, showing the relative life of wells and the number of live wells at the beginning of each year, illustrate this report, which sketches, in successive chapters, the history, boring, piping and refining of petroleum, with special accounts of the principal wells.

A map and profile section from Clarion to Butler, carefully made by Mr. Lucas, C. E., and kindly presented to the Survey by Mr. Jos. D. Potts, of Philadelphia, will be published in the same volume; and a reduction of Mr. Lucas' section to the same vertical and horizontal scale will show how exceedingly gentle is the south-westward slope of the whole oil country underground.

There was no good reason to expect from the survey, in the first four months of its progress, any discoveries of importance. Discoveries were the natural and parennial fruits of the First Survey, because nothing was really known about tne geology of the State, and little about that of the United States. Every step made then was made on virgin ground; every observation had the charm of novelty; every mountain and valley was a separate riddle to be solved; every formation was strange to the eye, its contents unpredicted, its strata unclassified, their outcrops untraced. The region was as much a field for discovery as are the territories of Colorado, Arizona, Utah and Nevada now.

All this is of the past. The light of thirty years has rested on the State, revealing to geologists every feature of its structure, seeming to leave no place for important discoveries. The main business of the present survey is therefore critical examination, analysis, measurement, description and publication; the precise determination of qualities and quantities; the finer estimation of place, position, posture, mineral nature, metallurgical fitness, and mining facility; the verification of what has been the subject of the comparatively rough statements of past reports, and their authoritative re-statement in a more detailed way, more intelligibly, and with all the corrections and additions which the business transactions of many individuals and incorporated companies enable us to make.

Questions however of the highest importance have, in fact, never yet been answered; in spite of the successful efforts of the first survey, and in spite of all the mining and exploring that has been done in the last thirty years. Every year actually reveals something unexpected. Discoveries are always possible; and some have been made in the four months of field operations just terminated.

Professor Prime seems to have found an important key to the brown hematite deposits of the limestone valleys, by connecting them (instrumentally) with certain hydro-mica slate beds containing potash, (damourite?) which have hitherto been over-looked. Faults discovered near Dillsburg, in York county, may explain the structure of the whole New Red part of south-eastern Pennsylvania; and this discovery affects the mining value of a large range of iron ore deposits.

The discovery of a great fault at Port Clinton, on the Schuylkill, by Mr. Chance, throws light on some perplexing portions of our geology.

The discovery of the extension of the outcrops of two small coal beds (already known to exist in *one* of the Allegheny Mountain ravines, near Altoona) makes an important addition to our systematic geology; for these beds lie several hundred feet below the Conglomerate, and may possibly be of workable thickness in neighborhoods where they are much wanted.

The identification of one of the "Mountain Sandrocks" of Venango county, by Mr. Carll, with the "Conglomerate" of Olean and Chattauqua Lake, puts an entirely new face on the geology of our northern and north-western counties, and on the published geologies of New York and Ohio; and will have an important influence on questions of underground measurement in the oil regions.

Such discoveries are sufficient to encourage the hope that the present survey of the State will not merely be successful in its proper work of gathering up and publishing a mass of useful discoveries made during the last twenty or thirty years, and never yet described and explained to the citizens of the Commonwealth, whose common property they ought to be, but may make new and equally important discoveries of its own. This can hardly fail to be the case, if the same care in observing all facts, and the same precise method of bringing them by careful instrumental measurements into their proper relations to each other, continue to govern the operations of the survey, which the reports, published this winter, will show to have been the rule in the first field season just closed.

It is a source of regret that several districts of the State have had to be entirely neglected for want of an appropriation equal to the original estimate of the working expenses of so large a survey. Two more working field parties might then have been employed.

SECOND GEOLOGICAL SURVEY OF PENNSYLVANIA.

Reports of the Progress of the Work in 1875, *made to the Governor and Board of Commissioners August* 4*th and November* 4*th, corrected to December* 31*st*, 1875.

Experience has proved that all surveys require equal amounts of field work and office work. In railway surveying a heavy force is usually kept at work in the office, plotting, mapping, sectioning, calculating, and draughting working plans, because the field work is pushed on through the winter months. But geological surveying is necessarily stopped by the first general heavy fall of snow, and cannot re-commence until the weather becomes settled between the first and fifteenth of May. The whole corps is therefore kept afoot seven months, and the remaining five months is employed in preparing reports, and illustrating these with drawings.

The winter of 1874–'5 was spent in this manner by the geologists of the survey and their aids. It was impossible to diminish the corps and get through the work. The discussion of geological note-books is peculiarly tedious, and requires much study of already published works. Geological maps cover large areas; and geological sections, unless accurately constructed and carefully drawn, are injurious rather than useful. The age of fancy pictures in geology has passed away, and the demands of the business world require geological statements to be as nearly as possible mathematically correct.

Sufficient office work was accomplished to permit the corps to take the field in May. The work of the laboratory, and the arrangement of the collections, went on uninterruptedly throughout the year.

By order of the Board of Commissioners two new districts were occupied; one in the south-west corner of the State, and one on the New York State line.

With the exception of two of the volunteer aids, the same parties, as in 1874, took the field in 1875, each in its own district.

Prof. Prime, Mr. Clark and Mr. Kent, continued the close survey of the Lehigh and Northampton limestone ore-belt; Prof. Frazer, Mr. Lehmann and Mr. Edwards, that of the York and Adams county ore-belts and the South Mountains; Mr. Dewees, that of the fossil ore-belts.

Mr. Billen and Mr. Ashburner made a special survey of the Aughwick valley and East Broad Top.

Mr. W. G. Platt was appointed aid to Mr. Franklin Platt, whose survey extended from Clearfield county southward to the State Line.

Mr. Sanders continued his mapping of Canoe valley and Morrison's cove. Mr. Young studied the coke manufacture, and surveyed the whole line of the Youghiogheny river.

Mr. Carll, with Mr. Hatch and Mr. Hale, continued the sur vey of outcrop and underlay of the Oil and Mountain Sands.

Mr. Wrigley's connection with the survey, having been for a special purpose, ceased on the handing in of his report.

Prof. J. J. Stevenson and his aid, Mr. J. C. White, was appointed to occupy a new district: Washington and Greene.

Mr. Andrew Sherwood, assisted by his brother, was appointed to survey another new district: Tioga and Bradford.

Mr. Chance made special studies of the Water Gaps, and lastly a special survey of the Beaver River waters.

Mr. M'Creath, and his efficient temporary assistant, Mr. Ford, have continued to analyze at Harrisburg the iron ores, coals, fire-clays and limestones sent in from the different districts.

Mr. C. E. Hall not only received and arranged, studied and named all fossils collected, but made excursions alone, or with volunteer students, to study in place characteristic species.

Dr. Genth has determined mineral species, and made, with some volunteer aid from students, numerous analyses of new minerals. He is chiefly occupied with special studies of the trap-rocks, and of the mineral spring waters of the State.

Dr. T. Sterry Hunt is preparing a special report on the geological connection of the trap-dykes, iron ores and azoic.

Dr. S. P. Sadtler has commenced an analytical study of the gas used in the western counties for iron making.

Mr. E. B. Harden has prepared a large number of drawings

of all kinds for publication in the Reports of Progress, and been unremitting in his attention to requisitions for materials and facilities made upon headquarters.

Mr. Charles Allen has consulted the railway superintendents and surveyors of the State, and collected from all quarters the materials for his report on the Levels of places in Pennsylvania.

I have myself done little beyond trying to make the survey efficient, and directing it along channels of practical utility to the Commonwealth, without sacrificing its scientific value. All classes of the community have shown a ready willingness to aid its progress, and many citizens of the State have exhibited a zealous interest in its success, by freely placing at its disposal their accumulations of facts, reports of private surveys, analyses, experiments and specimens. All these will be published in due time in the volumes of the Reports of Progress to which they belong. Some of them will be mentioned in the following list of reports already published, or to be published next year.

The field work of the year 1875 has been of three kinds, which explains a noticeable feature of the small skeleton maps of 1874, 1875, 1876, herewith appended, viz: the smallness of some of the districts surveyed, and the largeness of others.

The iron ore limestone areas of the State require and have received, and will continue to receive, the most minute and accurate surveying. Contour-line maps have been slowly and patiently made of these areas, for it is useless to attempt to hurry over work of this kind. The Oil Districts demand and receive the same kind of attention.

The bituminous coal areas, on the other hand, although in the end they ought to be surveyed minutely and mapped in contours, get for the present, and at first, only a preliminary survey of every known coal opening, with local sections. The geology is simple in structure; the rocks lie flat; and there are few faults. Two counties can, therefore, be thus surveyed by one party in one season. In 1874, Clearfield and Jefferson were thus surveyed. Cambria, Somerset, Washington, Greene, and parts of other counties in 1875. In 1876, Fayette, Westmoreland, Indiana, Armstrong, and other counties west of the Allegheny river will receive the same attention. Tioga and Bradford have been studied in like manner, although not in the coal region proper. Their coal fields will be reported upon separately.

The *size* of the tinted portions of the little maps showing the rate of progress of the survey is, therefore, no visible indication of the relative amount of work done by each party. Quite as much zeal and skill were lavished upon the *small* black spots as upon the *large* ones; but in quite a different way, owing to the different kinds of geological difficulties to be overcome, and the different results to be attained. And the same remark may be made of the reports.

The third kind of work consists of statistical studies connected with the geology of special districts, like that of the *Coke* region. A special service was necessary for tabulating all known *Levels* in the State, to which the surveying parties might refer. Special studies of mineral *springs*, *gas wells*, &c., cannot be shown upon a map, and cannot be brought into direct comparison with regard to the general field work.

It is a matter of constant surprise and anxiety how great a variety of demands can legitimately be made upon a geological survey, even when strictly confining itself to its proper business, and doing nothing for the sake of display. It is understood that, since the State has a Bureau of Statistics, the geological survey is not called upon to divert any of its energies into that channel. It is also understood that neither Botany nor Zoology nor Agriculture nor Metallurgy nor Manufactures form natural and necessary parts of a geological survey, but only come into view incidentally, when fossil plants or animals are fonnd in the rocks, or there is a question of the applicability of minerals as fertilizers, fuels, fluxes, ores, or stuffs for chemicals. Yet a geologist must keep his eyes open in all these directions, and suppress no fact because it is not *technically* geological, which bears upon these the most important business interests of the State.

The connection between Civil Engineering and Geology is so intimate that the engineer and geologist should mutually assist each other. Many geological questions arise in railway location and construction. On the other hand geological conclusions often depend on distances and levels, and on the shape of the surface, more than on the show of the rocks. All lines run by civil engineers come very useful to the geological survey, and are thankfully acknowledged and credited. All the documents preserved in the railway offices of the State have been freely placed at its disposal, and pains taken to give information necessary to make them useful to the survey. Companies have also not only readily offered their new materials for publication, but taken the trouble to prepare and present to the survey copies of railway survey maps, mineral land maps, plans of mines, diagrams of furnaces, &c., made in their own offices.

The Goldschmidt aneroids arrived from Zurich shortly before the commencement of field work, and were thoroughly tried. It cannot be said that their behavior in the peculiar service to which they were put proved satisfactory. They require the greatest care in handling, not for fear of breaking them, for they will bear rough usage, but to avoid large errors of reading. They are hard upon the eyes; awkward to carry and hold, and not smooth in their movement, on account of the cheap style in which they are put together. Their range is too great and their scale too small. It is easy to misread an elevation 20 feet too high or too low. One or two of the assistants, however, report satisfactorily of the accuracy of those they used; and doubtless there is a great difference in the manufacture.

Elliott & Co., of London, forwarded their copy of the Becker aneroid in May, and a thorough trial was given to it. The report was favorable. It is a good field-instrument for geological purposes. But it weighs twice as much as Becker's. Becker & Sons' establishment at Rotterdam, has been only recently broken up and transferred to New York, and the aneroids promised in the spring have not yet been manufactured. Enquiries for them have been received from other State Surveys and many private observers are impatiently looking forward to possessing them. It is of great importance to the ease and rapidity of the field work of 1876 that they should be made, for we are still dependent, in a great measure, upon our vertical circle running for covering intermediate areas with contour line work, and many of our geological indications are exhibited or suggested by the features of the surface, and these are oftentimes only to be expressed by accurate contour line representations.

The publication of the work of 1874, was commenced in the winter, and has been continued without intermission through the spring, summer and fall of 1875. Each assistant prepared his own report previous to taking the field in May, but the illustrative maps and drawings were slowly finished, photographed and printed at Philadelphia, while the manuscript text was composed and printed in the State Printing House at Harrisburg.

The Board has adopted the policy of—1. publishing results as fast as obtained; 2. publishing district reports separately; 3. publishing its own illustrations with each report; and 4. stereotyping for future use.

There can be no question about the propriety of this policy. But its advantages are not without a drawback. It imposes on the State Geologist unceasing labor as an editor, every day of the entire year. Every sentence of every report must be re-revised; the proper illustrations provided in advance; proofs and revises passed between author, editor and printer, artist and photo-lithographer; and much of the finishing and all of the final correcting done by the editor. In no other way can the work be carried on satisfactorily. Delays of various kinds are inevitable; and are sometimes unexpected and vexatious. The winter being spent in preparation of reports and illustrations, the printing must be done in the seven field-working months, while the assistants are scattered through the State. Although extra artist-work was employed, the need of accuracy in all delineations of surface and section work, and the need of legibility in the fac-simile reductions of the large originals by the photo-lithographic process, made the work slow. The weather is also an important element of uncertainty and delay in this process, to which, with all its drawbacks, modern science, as well as the practical arts, is so deeply indebted.

In all the illustrations of the reports of the Survey beauty of representation is kept secondary to precision of measurement and distinctness of subject. No attempt is made to show off the work by means of factitious or useless ornament. Whatever impedes the student from readily comprehending, or quickly referring to what he wishes to see, is sedulously avoided. Lettering and entitling is done in a plain and simple style; and the artistic harmony of the map lines is frequently but designedly marred for the sake of putting prominently before the eye those master lines which indicate the topography and geology of a district. As the courses of streams are always determined by the underground structure, special attention has been directed to making the principal drainage of a district evident at a glance. No road lines go upon a map beyond what are indispensable for indicating the localities which have been surveyed. Scales are multiplied and lengthened for easy use. Names are not allowed to accumulate, in order that the face of the map may be open and legible. Numbers are used instead of names, because they occupy less room; names and numbers being relegated to the margin of the map for reference. Small streams are left unnamed, because all the maps of the survey are intended to be used in the neighborhood; and the identity of brooks must be recognized by those who are familiar with the mines, furnaces, rock cliffs, &c., existing on or near them.

In like manner great pains have been taken with all vertical sections to keep them severely geological and practical. Letters are employed for the names of the formations; keys to the meaning of such letters being arranged in the margin. All sections are drawn strictly to nature, on the same scale vertical and horizontal; so measurements and calculations can safely be made by civil and mining engineers. Mere illustrative or diagrammatic sections are rejected, as being pretty but not valuable, except for primary education; and it is probable that our method of instruction is still defective at precisely this point. The schools would find their advantage in getting rid of the thousands of fancy pictures in science still in use, by substituting for them the real copies of nature which the recent work of men of science has furnished. It is intended that all the publications of the Pennsylvania survey shall be as real and as little fanciful as possible: show facts as they look; generalize to the least possible extent; and in a word, imitate the "working plans" of builders and artisans. Of course in this policy there is no arrogance; for it is simply the law of the scientific age we live in. While these publications must, of course, fall far short of a perfect obedience to it, yet in recognizing the law, and in endeavoring to be docile and obedient to it, whatever errors are committed may be set to the account of human infirmities, and not charged as reckless carelessness, or willful subserviency to traditional custom.

It is in the policy of the survey to make its reports of progress simple descriptions of work done, records of facts observed, and explanations of the local geology of each district within the limits of what is *known* by geologists; avoiding the discussion of abstruse questions, which do not concern the inhabitants, and are still subjects of speculation among geologists. If occasional references be made to those open questions, they are made under protest, and only for the purpose of awakening attention and stimulating enquiry. It is of the nature of reports of progress that they should sometimes err; but also that they should publish their own recantations. The world cannot wait for perfected knowledge of any subject of investigation. Progress presupposes improvement in abstract ideas, in the observation of facts, in the collection of statistics, and in the suggesting of explanations of the phenomena of nature. The careful and conscientious reports of trained observers must always possess a permanent value, in spite of the misstatements which lack of opportunity, defects of apparatus, or temporary bias of mind may allow to slip into them. And such observers will be first to acknowledge and earnest to redress such mistakes; while the blunders thus inadvertently committed will serve to evoke the zeal and make public the knowledge of others.

Such considerations have induced the Board of Commissioners of the Second Geological survey of Pennsylvania to issue annual reports of progress as fast as they can safely be prepared; and the Legislature having ordered that all the publications of the Survey shall be as cheap and handy as possible, and shall be sold at cost to all citizens, the following octavo volumes have been stereotyped and printed, and may be ordered at the rooms of the Board, No. 223 Market street, Harrisburg, either stitched or bound. Future reports will continue the paging; so that the series in any one district may hereafter be bound together as a separate volume.

A. First and Second Annual Reports of the operations of the

Survey in 1874 and 1875, preceded by a history of Geological Exploration in Pennsylvania and other States, with chapters on the Geological Nomenclatures in use; problems still to be solved, &c. By J. P. Lesley.

The first three chapters which are now published as forming a complete whole by themselves, were partly printed in December, 1874, and January 1875. But the pressing necessity for issuing the first reports of the Assistant Geologists in the winter, stopped the printing, which could not be resumed until November 1875. The same reason for postponing the publication of the remaining chapters now holds good. These will be issued in 1876. The text of the fourth chapter terminating at the bottom of page 200, will re-commence at the top of page 201. The index cannot be prepared until the whole is printed.

B. Preliminary report on the Mineralogy of Pennsylvania. By F. A. Genth. With an appendix on the Hydrocarbon Compounds. By Samuel P. Sadtler.—180 pages, with an index of 26 pages, and a map of the State (uncolored) for reference to the counties.

C. Report of Progress (in 1874) in York and Adams counties. By Persifer Frazer, Jr.—100 pages 8°, with two maps, ten sections, plates of microscopic rock slicings, &c.*

D. Report of Progress (in 1874) on the Brown Hematite ore ranges of Lehigh county, with a description of the mines lying between Emaus, Alburtis and Fogelsville. By Frederick Prime, Jr.—66 pages, and 8 pages of index, with a contour line map, and eight cuts. †

H. Report of Progress (in 1874) in the Bituminous Coal Fields of Clearfield, Jefferson, and parts of Centre and Armstrong counties. By Franklin Platt.—275 pages 8°, and 18 pages of index; eight maps, two sections and 139 cuts in the text.

I. Report of Progress (in 1874) in the Venango County Oil District. By John F. Carll. To which are appended: Observations on the Geology around Warren. By F. A. Randall.

---

* Ready for distribution in February, 1876.

† The printing of these cuts on the same press and on the same paper with the text, was a mistake. They are spoilt in the process. For future editions. impressions will be taken from the plates (which are excellent) on dry calendered paper and bound in.

Notes on the Comparative Geology of North-eastern Ohio, North-western Pennsylvania, and Western New York. By J. P. Lesley.—108 pages 8°, and 20 pages of index; two maps, a long section, and seven cuts in the text.

J. Special Report on the Petroleum of Pennsylvania, its production, transportation, manufacture and statistics.* By Henry E. Wrigley.—To which are added: Notes on a map and profile of a line of levels through the Butler, Armstrong and Clarion County Oil Fields; with a geological section from well-drillings. By D. Jones Lucas, Resident Engineer.—Notes on a map and profile of a line of levels along Slippery Rock Creek, with Leo Lesquereux's views of the connection between petroleum and fossil plants. By J. P. Lesley.—107 pages 8°, and 16 pages of index; five maps and sections, and a plate of the parts of an oil-rig and machinery.

M. Report of Progress (in 1874, 1875) in the Laboratory at Harrisburg. By Andrew S. McCreath.—93 pages 8°, and 12 pages of index.

The following volumes are in preparation and will be issued in all of 1876.

N. Report on Lists of 2,500 Levels collected in 1875. By Charles Allen. This will probably be published early in March, 1876, being intended as a first instalment of a complete record of all obtainable levels in and around Pennsylvania, for the use of the field parties of the Survey, for the use of Civil Engineers, and for the use of County Surveyors

B (Continued.) Dr. Genth's report of 1875.

C (Continued.) Prof. Frazer's report of 1875.

D (Continued.) Prof. Prime's report of 1875.

E. Report of Progress (in 1875) in Sinking and Canoe Valleys, and Morrison's Cove; Brown Hematite ores of the Lower Silurian valleys. By Mr. Platt. With maps by Mr. Sanders.

F. Report of Progress (in 1874, 1875) in the Juniata District. By John H. Dewees. This book has been delayed by the necessity for its copious illustration. It will give a description of all the fossil ore out-crops along Jack's, Shade, Blue Ridge, Black Log and Tuscarora mountains; with maps of Ferguson Valley, Logan Gap, Orbisonia and the East Broad Top Rail-

* Erratum, p. 62, line 4, for 2473.9 read 3617.8

road line; and elaborate cross sections, showing the Silurian and Devonian measures.

G. Report of Progress (in 1875) in Tioga and Bradford counties. By Andrew Sherwood. With a colored out-crop map.

H (Continued.) Mr. Platt's report of 1875, on Cambria and Somerset counties.

I (Continued.) Mr. Carll's report of 1875, on the Geology of the Oil Regions.

K. Report of Progress (in 1875) in Greene and Washington counties and parts of Beaver and Allegheny. By J. J. Stevenson. With cross sections, and a map showing the depths at which lie the principal coal beds.

O. Mr. Hall's report of the collections of fossils and rocks in 1874, 1875.

P. Mr. Lesquereux's report on the fossil Coal Plants of Pennsylvania and the United States; illustrated by 60 plates; and intended to serve as a hand-book for the Survey.

The three small maps of Pennsylvania accompanying this volume, are intended to exhibit roughly to the eye the amount of ground occupied by the field parties of the survey, in 1874; 2d, in 1874 and 1875 together; 3d, the amount of territory likely to be surveyed, in the various ways described in the preceding pages, in the three years 1874, 1875 and 1876. They show therefore the rate at which the survey advances under its present organization and with its present means.

It is evident from an inspection of these maps that nearly one-half of the surface of the State will remain to be examined in 1877, 1878 and 1879, if the Legislature see fit to order the continuance of the survey:—The important and difficult forest lands of the West Branch and Pine Creek;—the extremely complicated and difficult South-east section;—the whole Anthracite Region;—remaining portions of the Oil Region;—and numerous isolated small districts scattered about the State.

On the economical scale of $50,000 a year such a great work cannot be made to move on faster, nor would it be desirable to increase the scale of the expenditure; for good work ought not to be over-hurried. Its excellence depends upon attention to details, patient investigation of difficulties, the re-consideration of doubtful conclusions, and continual help from all quarters, which can be sought for only when needed, and never comes instantly at the call. The present organization of the survey is perfectly manageable, and its fruits not out of proportion to the publishing ability of the State Printing Office.

It must not be forgotten that an exact *topographical* survey of the entire State must be very much more tedious and expensive. But taking the county maps as we have them, and without contour-lines, we will be in a condition to issue some *geologically colored* county maps every year so long as the survey shall be permitted to go on, and be ready to publish such maps of them all, or a complete Geological Atlas of the State, at the close of 1879

www.ingramcontent.com/pod-product-compliance
Lightning Source LLC
LaVergne TN
LVHW050522100826
845148LV00002B/419

* 9 7 8 1 4 2 5 5 2 0 3 7 3 *